Recent Results in Cancer Research 97

Founding Editor
P. Rentchnick, Geneva

Managing Editors
Ch. Herfarth, Heidelberg · H. J. Senn, St. Gallen

Associate Editors
M. Baum, London · C. von Essen, Villigen
V. Diehl, Köln · W. Hitzig, Zürich
M. F. Rajewsky, Essen · C. Thomas, Marburg

Small Cell Lung Cancer

Edited by S. Seeber

With 44 Figures and 47 Tables

Springer-Verlag
Berlin Heidelberg New York Tokyo 1985

Professor Dr. Siegfried Seeber

Universitätsklinikum Essen, Innere Klinik und Poliklinik
(Tumorforschung) Westdeutsches Tumorzentrum
Hufelandstrasse 55, 4300 Essen, Federal Republic of Germany

Sponsored by the Swiss League against Cancer

ISBN-13:978-3-642-82374-9   e-ISBN-13:978-3-642-82372-5
DOI: 10.1007/978-3-642-82372-5

Library of Congress Cataloging in Puplication Data. Main entry under title: Small cell lung
cancer. (Recent results in cancer research; 97) Bibliography: p. Includes index. 1. Lungs
Cancer. I. Seeber, S. (Siegfried), 1941 –. II. Series. [DNLM: 1. Carcinoma, Oat Cell. 2.
Lung Neoplasms. W1 RE106P v. 97/WF 658 S6351] RC261.R35 vol. 97 616.99'4 s
[616.99'24] 84-20268 [RC280.L8]

© Springer-Verlag Berlin Heidelberg 1985
Softcover reprint of the hardcover 1st edition 1985

2125/3140–5 4 3 2 1 0

Contents

List of Contributors*

Arnold, H. 77[1]
Atay, Z. 25
Bleehen, N. M. 116
Drings, P. 87
Fiebig, H. H. 77
Greschuchna, D. 107
Gropp, C. 55, 65
Hansen, H. H. 47
Harper, P. G. 146
Havemann, K. 55, 65
Henß, H. 77
Holle, R. 65
Israel, L. 157
Jones, D. H. 116
Kaiser, D. 77
Koch, H. 77
König, R. 87

Lee, E. C. 37
Luster, W. 55, 65
Maassen, W. 107
Martinez, I. 107
Menne, R. 11
Müller, K.-M. 11
Neumann, H. A. 77
Niederle, N. 127
Schmähl, D. 1
Schütte, J. 127
Spang-Thomsen, M. 47
Souhami, R. L. 146
Vindeløv, L. 47
Vogt-Moykopf, I. 87
Whang-Peng, J. 37
Zeller, W. J. 1

---

* The address of the principal author is given on the first page of each contribution

1 Page on which contribution begins

Etiology of Small Cell Lung Carcinoma

W. J. Zeller and D. Schmähl

Deutsches Krebsforschungszentrum, Institut für Toxikologie und Chemotherapie,
Im Neuenheimer Feld 280, 6900 Heidelberg 1, Federal Republic of Germany

With regard to morphology, biology, and therapy it seems justified to consider small cell lung carcinoma (SCLC) separately from the other main forms of lung carcinoma (squamous cell, adenocarcinoma, large cell). This is expressed by the rough division of lung cancers into the small cell and non-small-cell groups. Regarding etiology, however, a separation of small cell carcinoma from the other types seems not to be justifiable and extremely difficult. Kreyberg (1962) has already combined squamous cell and small cell carcinomas into "group I tumors," pointing out that these types have common causative factors, especially environmental carcinogens, including tobacco smoking, occupational hazards, and ionizing radiation. In recent years the "unitarian" theory of the origin of lung cancers, suggesting a common stem cell, has been revived and is replacing the theory of a separate histogenesis of SCLC from the neural crest. Biochemical investigations have revealed that the L-dopa decarboxylase, the key enzyme of the so-called APUD tumors (Pearse 1969), can be detected not only in SCLC but, though at lower levels, in any morphological type of lung cancer (Baylin and Gazdar 1981). Both in lung cancer cultures and in patients (at relapse) a morphological conversion from SCLC to non-SCLC histology has been observed (Gazdar et al. 1981; Fer et al. 1983). It is therefore not surprising that most lung carcinogens are associated with more or less the entire spectrum of lung cancer types in man.

In a reasonably large number of studies no clear-cut data concerning the incidence of SCLC are available. To some extent this is due to the lack of histological standardization, especially in earlier reports. Also, the combination of SCLC with squamous cell carcinoma into group I tumors has led to the loss of some information on SCLC incidence. In recent years more detailed tumor typing has been pursued, which is indispensable for a therapeutic strategy.

The present survey on the etiology of SCLC is restricted mainly to some essential causative factors of this tumor type. For further information on the etiology of bronchogenic carcinoma, and in particular of SCLC, the reader is referred to the literature (Harris 1978; Greco et al. 1981; Zeller and Schmähl to be published).

The rate (number per 100,000/year) of SCLC varies with the time period in which the data were collected and in addition shows some regional differences. Annegers et al. (1978) calculated for the population of a county in Minnesota for the decades 1935–1954 a rate of 2.1/100,000/year in men, and for the decade 1965–1974 a rate of 6.0/100,000/year. Weiss (1981) estimated a rate of 8.8/100,000 man-years for men in Philadelphia for the period between 1951 and 1965. The relative frequency of SCLC is about 20% (squamous cell carcinoma 45%, adenocarcinoma 25%, and large cell carcinoma 10%) (Eckert et al. 1979; Katlic and Carter 1979; Hermanek and Gall 1979). Regarding the relative frequency of SCLC in different age groups, varying data can be found in the literature. Kennedy (1972)

Recent Results in Cancer Research. Vol. 97
© Springer-Verlag Berlin · Heidelberg 1985

reported on 40 cases of lung cancer in patients below the age of 40: 19 of 26 male (73%) and 7 of 11 female patients (64%) had SCLC. Putnam (1977), on the other hand, reported a predominance of adenocarcinomas (> 40%) and only 2 oat cell and 2 anaplastic carcinomas among 16 men and 8 women less than 40 years of age with lung cancer. Altogether, however, there is an indication that SCLC is commoner in younger age groups (Kreyberg 1969; Weiss 1981).

As specific etiologic factors of SCLC the essential known lung carcinogens can be enumerated: (a) smoking, (b) radiation, (c) asbestos, and (d) chemical carcinogens. The involvement of air pollution by PAH emission in the etiology of bronchogenic carcinoma, and in particular of SCLC, cannot be satisfactorily assessed at present.

Auerbach et al. (1961, 1962) demonstrated convincingly that tobacco smoke is one of the main factors in the development of human bronchogenic carcinoma. They recorded three principal types of epithelial changes as a consequence of cigarette smoking: increase in the number of cell rows (between the tunica propria and the surface row of ciliated columnar cells); loss of cilia; and presence of atypical cells. They found a high degree of correlation between the number of cigarettes smoked and the frequency of these changes. After cessation of smoking epithelial lesions are reduced (66.6% in sections from ex-smokers versus 97.8% in sections of current smokers). In the sections from ex-smokers atypical nuclei were found in 50% versus 100% for current smokers.

The causal relationship of cigarette smoking to lung cancer is further supported by studies among Seventh-Day Adventists, the majority of whom do not smoke: those dying of lung cancer in this population group almost always have a history of smoking (Lemon and Walden 1966).

Another convincing hint that cigarette smoke acts as a lung carcinogen is the observation that the age distribution of lung cancer manifestation shifted downwards as the number of cigarettes smoked per day increased and as the age at which smoking began decreased (Weiss 1973).

Auerbach et al. (1975) found that all cell types of bronchogenic carcinomas seemed to be related to smoking to about the same degree. A comparable result was found by Beamis et al. (1975) in a study of 1,145 patients; these authors concluded that all cell types are related to cigarette smoking.

Yesner et al. (1973) concluded that the number of cigarettes smoked was directly related to SCLC but not to squamous cell carcinoma or other types. In a 10-year prospective study of 6,136 men, Weiss et al. (1972) found a dose-response relationship between smoking and well-differentiated squamous cell carcinoma, small cell carcinoma, and adenocarcinoma, whereas in the poorly differentiated squamous cell carcinoma no such correlation was found. In a study in 1,682 patients between 1962 and 1975 Vincent et al. (1977) found that regardless of the histopathology over 85% of all patients with lung cancer had been smokers. They were unable to disassociate smoking as a causative factor from any lung cancer type; among nonsmokers on the other hand they found a prevalence of adenocarcinoma, a result in line with the data of other investigators (Yesner et al. 1973).

Stayner and Wegman (1982) reported in a case-control study with access to the Third National Cancer Survey that cigarette smoking was significantly associated with squamous cell, small cell, and adenocarcinoma, and that the relationship with SCLC was strongest. In the aforementioned study of Kennedy (1972) in 40 lung cancer patients below the age of 40, two-thirds of whom had SCLC, only 4 of the patients were nonsmokers.

It is striking that in women with a known smoking history a distinct increase in squamous cell and small cell carcinomas can be observed (Beamis et al. 1975; Chan et al. 1979). In 1960 the ratio of male to female lung cancer death rates reached a peak, and since then it has been declining as result of this rapid increase in lung cancer in women (Burbank 1972). Altogether the data suggesting that smoking is a major cause of small cell lung cancer are convincing.

A further essential causative factor in human lung cancer is ionizing radiation. Investigations among uranium miners showed an increase in squamous cell, small cell, and adenocarcinoma due to inhalation of radioactive substances; the incidence of small cell undifferentiated carcinoma was increased to the greatest extent (Archer et al. 1974) (Table 1). Horáček et al. (1977) observed an increase in the frequency of small cell and squamous cell carcinomas among uranium miners; they observed no increase in the frequency of adenocarcinomas. Archer et al. (1976) observed a synergistic action between smoking and radiation exposure. The lung cancer rate in heavy smokers who had heavy radiation exposure was about 10 times the rate in nonsmokers with heavy radiation exposure, but 67 times the rate in nonsmokers with low radiation exposure.

With regard to this synergistic action of ionizing radiation and tobacco smoke in the etiology of human lung cancer, the low concentrations of polonium-210 (^{210}Po) and of lead-210 (^{210}Pb) (the parent of ^{210}Po) in inhaled mainstream smoke could also be contributory factors for the development of bronchogenic carcinomas (Radford and Hunt 1964; Martell 1975). Whole-body irradiation (Ishimaru et al. 1975) and therapeutic X-ray exposure (Court-Brown and Doll 1965) are also associated with the risk of lung cancer.

Among respiratory carcinogens asbestos plays an important role (Doll 1955). The annual world production of asbestos amounts to about 4.2 million tons (1983) and the large number of persons at risk from exposure to this mineral is of particular concern. Asbestos increases the risk both of bronchogenic carcinoma and of pleural mesothelioma. With regard to the development of bronchogenic carcinoma a striking synergism between asbestos exposure and cigarette smoking has been observed; the development of mesotheliomas after asbestos exposure is apparently not influenced by cigarette smoke (Wagner et al. 1971). Selikoff et al. (1968, 1980) demonstrated that cigarette smokers who work with asbestos are about 92 times more likely to die of bronchogenic carcinoma than those who neither smoke nor are exposed to asbestos. Asbestos workers who stop smoking have a declining risk of lung cancer compared with those who continue smoking (Hammond et al. 1979). The observations that lung cancer is almost as rare in nonsmoking asbestos workers as in those not exposed to asbestos and that asbestos-associated lung cancer is almost entirely found in smokers have raised the question as to whether asbestos is merely a mediator by which the tobacco effect is enhanced (Kannerstein and Churg 1972).

With regard to the histological types of lung cancer after asbestos exposure, Kannerstein and Churg (1972) observed no differences between an asbestos-associated group of 50 patients and a control group. SCLC occurred in 6 of the 50 asbestos-associated cases and in 8 of the 50 control cases (Table 2). Whitwell et al. (1974) found no difference in the frequency of SCLC between patients with less severe asbestosis and those with moderate and severe asbestosis (Table 3). They found an increase in the frequency of adenocarcinomas only from 25% to 38%; however, this difference was not statistically significant.

In Tables 4 and 5 some further compounds with proven carcinogenicity for human lung tissue are documented. Table 5 shows the frequency of histological types of lung cancer after exposure to some chemical agents. It is apparent that inhaled occupational lung carcinogens produce their characteristic frequency pattern including all histological types, so that a separation of adenocarcinomas as group II tumors that according to Kreyberg's hypothesis (Kreyberg 1962) are not caused by inhaled carcinogens does not seem to be

Table 1. Observed and expected bronchogenic carcinomas among U.S. uranium miners, 1950 – 1970. (Archer et al. 1974)

Radiation dose WLM[a]	Squamous cell			Small cell			Adenocarcinoma			Large cell or other		
	Observed no.	Expected no.	Ratio	Observed no.	Expected no.	Ratio	Observed no.	Expected no.	Ratio	Observed no.	Expected no.	Ratio
1– 119	0	1.25	0	2	0.30	6.67	0	0.28	0	0	0.30	0
120– 359	4	1.81	2.21	7	0.43	16.28	2	0.41	4.88	1	0.43	2.33
360– 839	6	2.10	2.86	10	0.50	20.00	0	0.47	0	1	0.50	2.00
840–1799	8	1.79	4.47	13	0.43	30.23	1	0.40	2.50	0	0.42	0
1800–3319	8	1.02	7.84	22	0.24	91.67	2	0.23	8.70	0	0.24	0
≥ 3320	4	0.30	13.33	12	0.07	171.43	3	0.07	42.86	1	0.08	12.5
Total	30	8.27	3.63	66	1.97	33.50	8	1.86	4.30	3	1.97	1.52

[a] Working level month

Table 2. Cell types of bronchogenic carcinoma associated with asbestos exposure. (Kannerstein and Churg 1972)

	Squamous cell	Anaplastic small cell	Adeno-carcinoma	Anaplastic large cell	Combined	Unclassified
Asbestos-associated Subjects	11	11	11	6	8	3
Controls	12	14	9	8	7	0

Table 3. Histological type of tumor in 86 cases, graded by severity of asbestosis. (Whitwell et al. 1974)

Cell type	Normal lung and mild asbestosis		Moderate and severe asbestosis	
	No.	%	No.	%
Squamos cell	8	28.6	11	19.0
Small cell	7	25.0	16	27.6
Adenocarcinoma	7	25.0	22	37.9
Other	6	21.4	9	15.5

justified. Although the number of cases in some studies is low, for most compounds a distinct association between exposure and the development of SCLC can be noted. The percentage of SCLC varies from 0 (vinyl chloride) to 74 (chloromethylethers). For the remaining compounds the percentage of SCLC lies between 15 and 33. The predominance of SCLC after chloromethylether (CME) exposure is noted in several reports and suggests that SCLC is a specific response to inhalation of these compounds [especially bis(chloromethyl)ether] (Figueroa et al. 1973; Thiess et al. 1973; Lemen et al. 1976a; Weiss et al. 1979). Table 6 shows the distribution of 43 cases of lung cancer by cumulative exposure to CME and by histological type. After moderate and heavy exposure the proportion of SCLC was $\geq$ 80%. Another essential observation was that 5 of the 20 cases in this study with moderate and heavy exposure were nonsmokers and that the age at which lung cancer was diagnosed in these two groups was considerable lower than in the other groups (Weiss et al. 1979). While Lemen et al. (1976a) concluded from their data that cigarette smoke might interact with CME exposure in a synergistic fashion, Weiss (1980) reported a higher risk of developing lung cancer in men who were not smoking: in a prospective epidemiological study of 125 workers 11 developed lung cancer; 6 cases were observed in 13 nonsmokers and ex-smokers and 5 in 38 current smokers. This inverse relationship between lung cancer risk and cigarette smoking in CME workers is in contrast to observations made in asbestos workers and in uranium miners. One possible explanation is that the carcinogenic effect of CME may be neutralized to some extent in smokers (Weiss 1980).

It is remarkable that in experimental animals (rats) after inhalation of bis(chloromethyl)ether the predominant histological cell type of lung carcinomas observed was squamous cell carcinoma; small cell carcinomas were not observed in rats. A single undifferentiated carcinoma of the lung was seen in 1 of 100 hamsters that died at 501 days after 334 exposures (Kuschner et al. 1975), thus confirming that the observation of small cell carcinomas is a rarity in experimental animals (Nettesheim et al. 1970; Karbe and Park 1974). After chronic intratracheal instillation of benzo(a)pyrene in 347 Syrian golden hamsters we observed 44 carcinomas of the respiratory tract; 5 of them (11%) were adenocarcinomas and the remaining cases were squamous cell carcinomas; no small cell carcinomas were observed (W. J. Zeller et al., in press). Blair (1974), on the other hand, reported a higher incidence of SCLC in experimental animals. He observed SCLC in 20 of 100 Sprague-Dawley rats after intratracheal instillations of benzo(a)pyrene with ferric oxide as carrier dust; 18 of these 20 rats, however, also exhibited squamous cell and adenocarcinomas in other areas of the lung.

Table 4. Occupational respiratory carcinogens.[Adapted from Wynder and Hecht (1976) and Frank (1978)]

Carcinogen	Latent period (years)		Approximate relative risk	Occupational groups
	Average	Range		
Arsenic	25	10 – 56	2 – 7	Smelter men, vineyard workers, sheep dip manufacturers
Asbestos		20 – 30	1.5 – 12	Insulation workers, shipyard workers
Chloromethylethers		10 – 20	7 – 24	Chloromethylether production workers
Chromium	24	3 – 58	3 – 15	Chromium ore processing pigment manufacturers
Carbon compounds including coke		15 – 20	1.8 – 3.5	Coke oven workers, gas workers, roofers, rubber workers
Mustard gas	20	10 – 27	2 – 36	Mustard gas production workers
Nickel	20	9 – 27	5 – 10	Nickel refinery workers
Radiation	25	10 – 45	1.7 – 29	Uranium miners, hard-rock miners

Table 5. Chemical agents and histological types of lung cancer. (Modified from Weiss 1981)

Chemical agent	No. of cases	Squamous cell		Small cell		Adeno-carcinoma		Large cell or other		References
		No.	%	No.	%	No.	%	No.	%	
Vinyl chloride	8	–	–	–	–	3	38	5	63	Waxweiler et al. 1976
Nickel	39	26	67	6	15	7	18	–	–	Pedersen et al. 1973; Kreyberg 1978
Cadmium	8	3	38	2	25	–	–	3	38	Lemen et al. 1976b
Arsenic	60	21	35	16	27	18	30	5	8	Axelson et al. 1978; Wicks et al. 1981
Chromate	18	13	72	5	28	–	–	–	–	Abe et al 1982
Acrylonitrile	6	4	67	2	33	–	–	–	–	O'Berg 1982
Chloromethylethers	47	2	4	35	74	5	11	5	11	Figueroa et al. 1973; Lemen et al. 1976a; Weiss et al. 1979

Table 6. Distribution of lung cancer cases in chemical workers by histological type and cumulative exposure to chloromethyl ethers. (Weiss et al. 1979)

Exposure	Cases (no)	Histologic type				
		Squamous cell		Small cell		
		No.	%	No.	%	
O	15	6	40	3	20	
Light	8	1	13	2	25	
Moderate	10			9	90	
Heavy	10			8	80	

Exposure	Adenocarcinoma		Large cell		Other	
	No.	%	No.	%	No.	%
O	3	20	3	20		
Light	4	50	1	13		
Moderate	1	10				
Heavy			1	10	1	10

Table 7. Association of lung cancer types with scars, (Auerbach et al. 1979)

Histological type	Number of cases of lung cancer	Presence of scar	
		No.	Percent of total
Squamous cell	442	15	3.4
Small cell	246	0	–
Adenocarcinoma	295	59	20.0
Large cell	195	8	4.1
Mixed type	8	0	–
Total	1,186	82	6.9

Finally, in the discussion of the etiology of SCLC it must be pointed out that there is evidently no association between SCLC and lung scars. The majority of scar cancers of the lung are adenocarcinomas (Lüders and Themel 1954). Although other histological types can also be observed in patients with scars (Eck et al. 1969), it is accepted that scars play no decisive role in the etiology of SCLC. Table 7 gives the result of a review of 1,186 cases of lung cancer among 7,629 autopsied cases over a 21-year period by Auerbach et al. (1979). Of these cancers 82 were related to scars, and it is noteworthy that none of these was of the small cell type.

References

Abe S, Ohsaki Y, Kimura K, Tsuneta Y, Mikami H, Murao M (1982) Chromate lung cancer with special reference to its cell type and relation to the manufacturing process. Cancer 49: 783−787

Annegers JF, Carr DT, Woolner LB, Kurland LT (1978) Incidence trend and outcome of bronchogenic carcinoma in Olmsted County, Minnesota, 1935−1974. Mayo Clin Proc 53: 432−436

Archer VE, Saccomanno G, Jones JH (1974) Frequency of different histologic types of bronchogenic carcinoma as related to radiation exposure. Cancer 34: 2056−2060

Archer VE, Gillam JD, Wagoner JK (1976) Respiratory disease mortality among uranium miners. Ann NY Acad Sci 271: 280−293

Auerbach O, Garfinkel L, Parks VR (1975) Histologic type of lung cancer in relation to smoking habits, year of diagnosis and sites of metastases. Chest 67: 382−387

Auerbach O, Stout AP, Hammond EC, Garfinkel L (1961) Changes in bronchial epithelium in relation to cigarette smoking and in relation to lung cancer. N Engl J Med 265: 253−267

Auerbach O, Stout AP, Hammond EC, Garfinkel L (1962) Bronchial epithelium in former smokers. N Engl J Med 267: 119−125

Auerbach O, Garfinkel L, Parks VR (1979) Scar cancer of the lung. Cancer 43: 636−642

Axelson O, Dahlgren E, Jansson CD, Rehnlund SO (1978) Arsenic exposure and mortality: a case-referent study from a Swedish copper smelter. Br J Ind Med 35: 8−15

Baylin SB, Gazdar AF (1981) Endocrine biochemistry in the spectrum of human lung cancer: Implications for the cellular origin of small cell carcinoma. In: Greco FA, Oldham RK, Bunn PA Jr (eds) Small cell lung cancer. Grune and Stratton, New York, pp 123−143

Beamis JF Jr, Stein A, Andrews JL Jr (1975) Changing epidemiology of lung cancer. Med Clin North Am 59: 315−325

Berge T, Toremalm NG (1975) Bronchial cancer − a clinical and pathological study. Scand J Respir Dis 56: 120−126

Blair WH (1974) Chemical induction of lung carcinomas in rats. In: Karbe E, Park JF (eds) Experimental lung cancer, carcinogenesis and bioassays. Springer, Berlin Heidelberg New York, pp 199−206

Burbank F (1972) U.S. lung cancer death rates begin to rise proportionately more rapidly for females than for males: a dose-response effect? J Chronic Dis 25: 473−479

Chan WC, Colbourne MJ, Fung SC, Ho HC (1979) Bronchial cancer in Hong Kong 1976−1977. Br J Cancer 39: 182−192

Court Brown WM, Doll R (1965) Mortality from cancer and other causes after radiotherapy for ankylosing spondylitis. Br Med J II: 1327−1332

Doll R (1955) Mortality from lung cancer in asbestos workers. Br J Ind Med 12: 81−86

Eck H, Haupt R, Rothe G (1969) Die gut- und bösartigen Lungengeschwülste. In: Uehlinger E (ed) Atmungswege und Lungen. Springer, Berlin Heidelberg New York (Handbuch der speziellen pathologischen Anatomie und Histologie, vol 3/4)

Eckert M, Hammann J, Höhn D, Schultess F (1979) Bronchialkarzinom. Diagnostik, Therapie und Ergebnisse. Fortschr Med 97: 1047−1050

Fer MF, Grosh WW, Greco FA (1983) Morphologic changes in small cell lung cancer. In: Greco FA (ed) Biology and management of lung cancer. Martinus Nijhoff, Boston, pp 109−124

Figueroa WG, Raszkowski R, Weiss W (1973) Lung cancer in chloromethyl methyl ether workers. N Engl J Med 288: 1096−1097

Frank AL (1978) Occupational cancer. In: Harris CC (ed) Pathogenesis and therapy of lung cancer. Marcel Dekker, New York Basel, pp 25−51

Gazdar AF, Carney DN, Guccion JG, Baylin SB (1981) Small cell carcinoma of the lung: Cellular origin and relationship to other pulmonary tumors. In: Greco FA, Oldham RK, Bunn PA Jr (eds) Small cell lung cancer. Grune and Stratton, New York, pp 145−175

Greco FA, Oldham RK, Bunn PA Jr (eds) (1981) Small cell lung cancer. Grune and Stratton, New York

Hammond EC, Selikoff IJ, Seidman H (1979) Asbestos exposure, cigarette smoking and death rates. Ann NY Acad Sci 330: 473–490

Harris CC (ed) (1978) Pathogenesis and therapy of lung cancer. Marcel Dekker, New York

Hermanek P, Gall FP (1979) Lungentumoren. Gerhard Witzstrock, Baden-Baden

Horáček J, Plaček V, Ševc J (1977) Histologic types of bronchogenic cancer in relation to different conditions of radiation exposure. Cancer 40: 832–835

Ishimaru T, Cihak RW, Land CE, Steer A, Yamada A (1975) Lung cancer at autopsy in A-bomb survivors and controls, Hiroshima and Nagasaki 1961–1970. Cancer 36: 1723–1728

Kannerstein M, Churg J (1972) Pathology of carcinoma of the lung associated with asbestos exposure. Cancer 30: 14–21

Karbe E, Park JF (eds) (1974) Experimental lung cancer, carcinogenesis and bioassays. Springer, Berlin Heidelberg New York

Katlic M, Carter D (1979) Prognostic implication of histology, size, and location of primary tumors. In: Muggia FM, Rozencweig M (eds) Lung cancer: progress in therapeutic research. Raven, New York, pp 143–150

Kennedy A (1972) Lung cancer in young adults. Br J Dis Chest 66: 147–154

Kreyberg L (1962) Histological lung cancer types: A morphological and biological correlation. Norwegian Universities Press, Oslo

Kreyberg L (1969) Aetiology of lung cancer: A morphological epidemiological and experimental analysis. Universitetsforlaget, Oslo

Kreyberg L (1978) Lung cancer in workers in a nickel refinery. Br J Ind Med 35: 109–116

Kreyberg L, Saxén E (1961) A comparison of lung tumor types in Finland and Norway. Br J Cancer 15: 211–214

Kuschner M, Laskin S, Drew RT, Cappiello V, Nelson N (1975) Inhalation carcinogenicity of alpha halo ethers. Arch Environ Health 30: 73–77

Lemen RA, Johnson WM, Wagoner JK, Archer VE, Saccomanno G (1976a) Cytologic observations and cancer incidence following exposure to BCME. Ann NY Acad Sci 271: 72–80

Lemen RA, Lee JS, Wagoner JK, Blejer HP (1976b) Cancer mortality among cadmium production workers. Ann NY Acad Sci 271: 273–279

Lemon FR, Walden RT (1966) Death from respiratory system disease among Seventh-Day Adventist Men. JAMA 198: 117–126

Lüders CJ, Themel KG (1954) Die Narbenkrebse der Lungen als Beitrag zur Pathogenese des peripheren Lungencarcinoms. Virchows Arch [Pathol Anat] 325: 499–551

Martell EA (1975) Tobacco radioactivity and cancer in smokers. Am Sci 63: 404–412

Nettesheim P, Hanna MG Jr, Deatherage JW Jr (eds) (1970) Morphology of experimental respiratory carcinogenesis. US Atomic Energy Commission (AEC Symposium Series 21)

O'Berg MT (1980) Epidemiologic study of workers exposed to acrylonitrile. J Occup Med 22: 245–252

Pearse AGE (1969) The cytochemistry and ultrastructure of polypeptide hormone producing cells of the APUD series and the embryologic, physiologic and pathologic implications of the concept. J Histochem Cytochem 17: 303–313

Pedersen E, Høgetveit AC, Andersen A (1973) Cancer of respiratory organs among workers at a nickel refinery in Norway. Int J Cancer 12: 32–41

Putnam JS (1977) Lung carcinoma in young adults. JAMA 238: 35–36

Radford EP Jr, Hunt VR (1964) Polonium-210: A volatile radioelement in cigarettes. Science 143: 247–249

Selikoff IJ, Hammond EC, Churg J (1968) Asbestos exposure, smoking and neoplasia. JAMA 204: 104–110

Selikoff IJ, Seidman H, Hammond EC (1980) Mortality effects of cigarette smoking among Amosite asbestos factory workers. JNCI 65: 507–513

Stayner LT, Wegman DH (1982) Smoking, occupation, and histopathology of lung cancer: A case-control study with the use of the 3rd national cancer survey. JNCI 70: 421–426

Thiess AM, Hey W, Zeller H (1973) Zur Toxikologie von Dichlordimethyläther-Verdacht auf kanzerogene Wirkung auch beim Menschen. Zentralbl Arbeitsmed Arbeitsschutz Prophyl Ergonomie 23: 97–102

Vincent RG, Pickren JW, Lane WW, Bross I, Takita H, Houten L, Gutierrez AC, Rzepka T (1977) The changing histopathology of lung cancer. Cancer 39: 1647–1655

Wagner JC, Gilson JC, Berry G, Timbrell V (1971) Epidemiology of asbestos cancers. Br Med Bull 27: 71–76

Waxweiler RJ, Stringer W, Wagoner JK, Jones J (1976) Neoplastic risk among workers exposed to vinyl chloride. Ann NY Acad Sci 271: 40–48

Weiss W (1973) Cigarette smoke as a carcinogen. Am Rev Respir Dis 108: 364–366

Weiss W (1980) The cigarette factor in lung cancer due to chloromethyl ethers. J Occup Med 22: 527–529

Weiss W (1981) Small cell carcinoma of the lung: Epidemiology and etiology. In: Greco FA, Oldham RK, Bunn PA Jr (eds) Small cell lung cancer. Grune and Stratton, New York, pp 1–34

Weiss W, Boucot KR, Seidman H, Carnahan WJ (1972) Risk of lung cancer according to histologic type and cigarette dosage. JAMA 222: 799–801

Weiss W, Moser RL, Auerbach O (1979) Lung cancer in chloromethyl ether workers. Am Rev Respir Dis 120: 1031–1037

Whitwell F, Newhouse ML, Bennett DR (1974) A study of the histological cell types of lung cancer in workers suffering from asbestosis in the United Kingdom. Br J Ind Med 31: 298–303

Wicks MJ, Archer VE, Auerbach O, Kuschner M (1981) Arsenic exposure in a copper smelter as related to histological type of lung cancer. Am J Ind Med 2: 25–31

Wynder EL, Hecht S (eds.) (1976) Lung cancer, UICC. UICC, Geneva (Technical report series, vol 25)

Yesner R, Gelfman NA, Feinstein AR (1973) A reappraisal of histopathology in lung cancer and correlation of cell types with antecedent cigarette smoking. Am Rev Respir Dis 107: 790–797

Zeller WJ, Schmähl D (1985) Ätiologie des Bronchialkarzinoms. In: Trendelenburg F (ed) Neoplasmen der Bronchien und der Lunge. Springer, Berlin Heidelberg New York (Handbuch der inneren Medizin, vol 4/4)

Zeller WJ, Schmähl D, Ivankovic S (1985) Inhalationsexperimente an Syrischen Goldhamstern: Kombination von chronischer Zigarettenrauchinhalation und intratrachealer Instillation von Benzo(a)pyren; Detoxifizierung des Zigarettenrauches durch Kohlefilter. Prax Klin Pneumol (in press)

Small Cell Carcinoma of the Lung: Pathological Anatomy

K.-M. Müller and R. Menne

Berufsgenossenschaftliche Krankenanstalten "Bergmannsheil Bochum", Universitätsklinik, Institut für Pathologie, Hunscheidtstrasse 1, 4630 Bochum 1, Federal Republic of Germany

Introduction

The demarcation of SCCL among malignant lung tumors is based on the light microscopic finding of relatively small tumor cells (Barnard 1926). The tumor cells of this tumor group, with sizes of 9 ± 1 µm and an average nucleus size of 7 ± 1 µm, are substantially smaller than the cells of squamous cell carcinomas, with a cell diameter of 16 ± 2.5 µm, and large cell carcinomas, with cell sizes of 40 ± 14 µm and nucleus sizes of 24 ± 6 µm (Brämer 1984; Fig. 1 and Table 1).

Data on the frequency of SCCL vary substantially. Most of the rates quoted are between 15% and 20% of all bronchial carcinomas. When referred exclusively to autopsies, SCCL accounts for up to 40% of all cases of lung cancer (Eck et al. 1969; Müller 1976). Reports based on surgical and autopsy material give about 18% SCCL (Fasske 1970; Hoppe 1974; Spencer 1977) and data from biopsy material, up to 30% of all lung cancer (Blaha 1983).

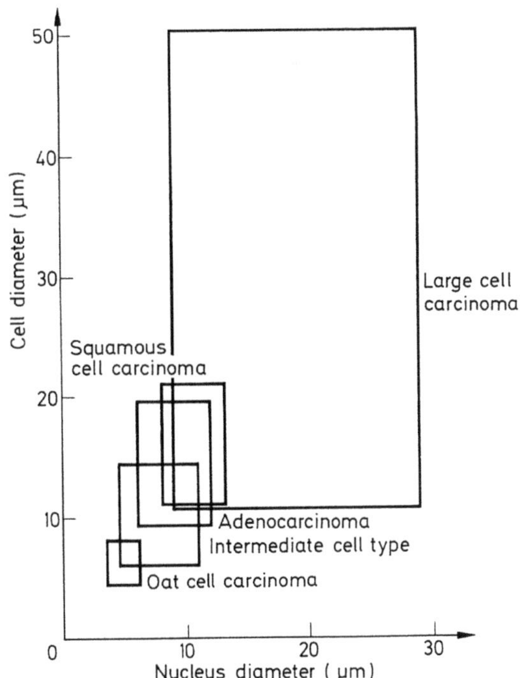

Fig. 1. Nucleus and cell diameters of different types of bronchial carcinoma; 2-sigma areas containing 95% of measured values (Brämer 1984)

Recent Results in Cancer Research. Vol. 97
© Springer-Verlag Berlin · Heidelberg 1985

Table 1. Average diameters of nuclei and cells in different types of bronchial carcinoma, together with frequency[a] and sex ratio[b] for each

Tumor type	Nucleus ⌀ (µm)	Cell ⌀ (µm)	Nucleus proportion (%)	Frequency (%)	Sex ratio male : female
Squamous cell carcinoma	9.2 ± 1.4	13.7 ± 2.4	68 ± 8	35 − 45	7 : 1
Oat cell carcinoma	5.1 ± 0.8	6.6 ± 1.2	81 ± 10		
Intermediate cell type	7.5 ± 1.3	9.8 ± 1.9	78 ± 10	15 − 25	4 : 1
Adenocarcinoma	8.5 ± 1.3	13.2 ± 2.3	65 ± 8	13 − 23	1 : 6
Bronchoalveolar carcinoma	7.5 ± 1.0	12.1 ± 2.0	62 ± 8	1.6 − 2.4	1 : 1
Giant cell carcinoma	16.4 ± 4.3	26,5 ± 8.9	65 ± 13		
Light cell carcinoma	8.3 ± 1.3	20.8 ± 3.9	41 ± 8	14.2 − 19	

[a] Data from Müller (1980)
[b] Data from Hackl (1973) and Müller (1976)

SCCL is characterized by particularly malignant and destructive growth and tends to extensive necroses. This type of lung cancer penetrates early into lymph and blood vessels. At the time of diagnosis, metastases in liver, brain, bones, and the suprarenal glands are frequently present (Fig. 2).

Macroscopic Findings

SCCL develops predominantly in the central and intermediate sections of the bronchial system. Central tumors occur more frequently than carcinomas primarily developed at the periphery, ratios of 2 : 1 and 3.5 : 1 being cited by Haupt and Stolper (1968) and Matthews (1979), respectively.
Only in exceptional cases is detection of the defined tumor starting point possible. In general, the diagnosis is made when the tumor is already at an advanced stage (Fig. 3A). Due to the fast growth rate, SCCL spreads early and rapidly in the bronchial mucosa and progresses peribronchially and perivascularly in hilipetal and hilifugal directions. Compared with squamous cell carcinomas the circumscribed nodular tumor is found in only about 10% of SCCL.
Due to the early and rapid intramural and perivascular tumor spread, bronchomediastinal lymph node metastases appear even in the early stages of the disease (Hackl 1969; Greschuchna and Maassen 1973).
The invasion of blood vessels even by very small primary tumors explains the precocious formation of hematogenous metastases. The peribronchial and perivascular centrifugal tumor spread at the boundary (Fig. 3B) explains a relatively premature pleural involvement even in the case of a centrally located SCCL, and reflects the development of pleural effusions. In recent studies, the frequency of pleural involvement in the case of central tumor localization is quoted at 13% only (Shimosato et al. 1982).
Reviews on this aspect have been published by Giese (1960), Eck et al. (1969), Schulze (1974), Spencer (1977), Shimosato et al. (1982), and Müller (1983).

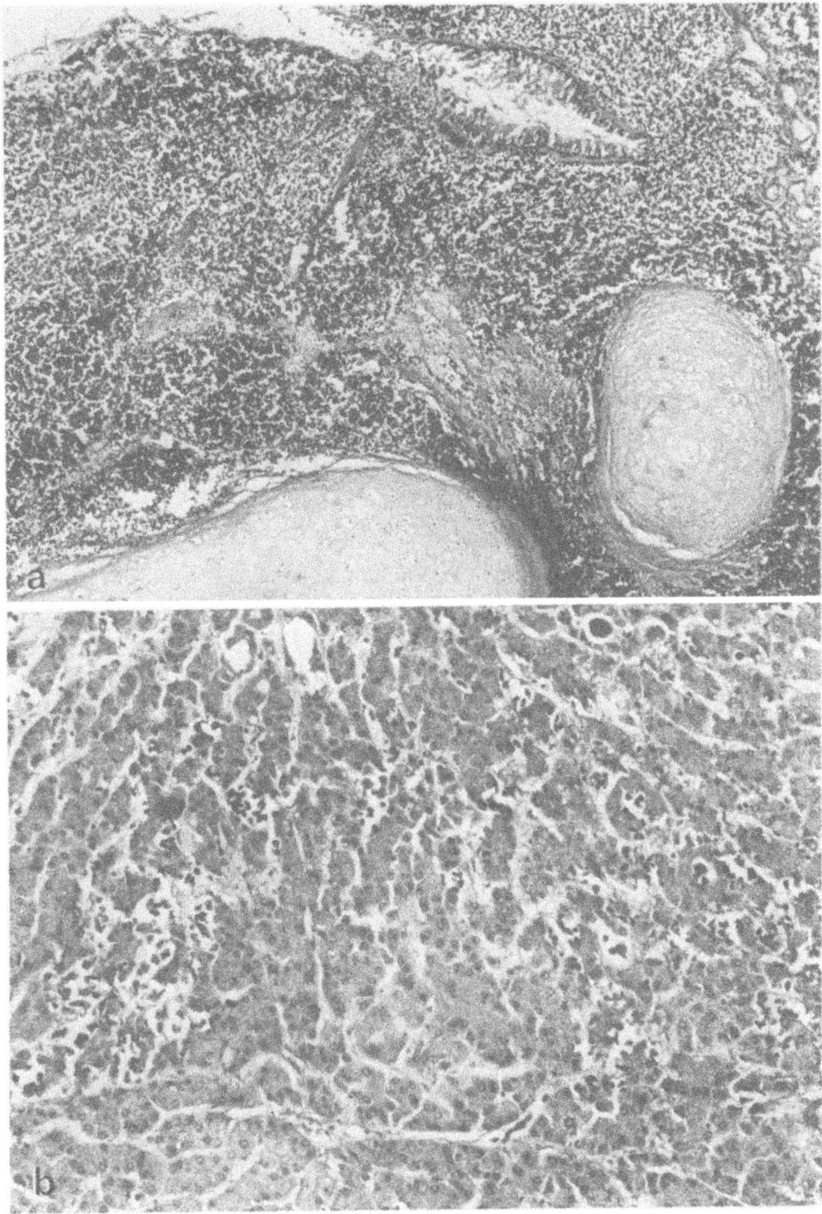

Fig. 2a, b. Small cell carcinoma of the lung (SCCL): **a** Massive diffuse mucosal infiltration proliferating into deep bronchial wall portions; **b** liver metastasis in the same patient with leukemoid infiltration of sinuses by tumor cells

Histological Classification and Histogenetic Aspects

Since differentiated histological evaluations of malignant lung tumors became possible, in almost every classification system suggested SCCL have been categorized as an individual tumor group. While Marchesani in 1924 differentiated between so-called basal cell cancer

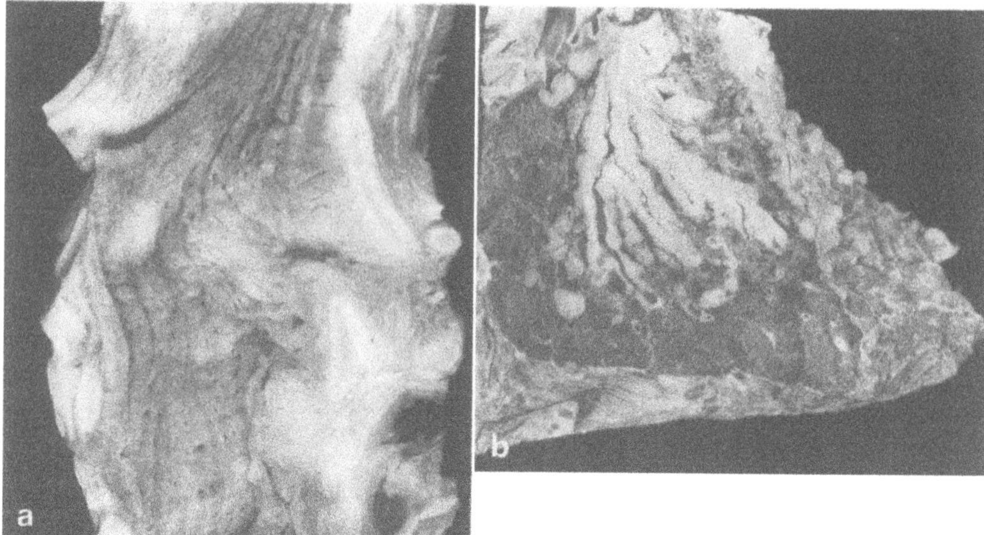

Fig. 3a, b. SCCL, macroscopic findings: **a** Stenosing plaque-like SCCL, localized in a subsegment bronchus. **b** finger-shaped, hilifugal, peribronchovascular tumor spread

and polymorphic squamous, and cylinder cell cancer types, the "primarily small cell carcinoma" appeared in the classification proposed by Fischer (1931). In the system used by Kreyberg (1962) small cell anaplastic carcinoma and squamous cell carcinoma form group 1. This group differs from group 2, which includes adenocarcinomas and carcinoids.

In every recent histological tumor classification, including the revised WHO classification of 1977, SCCL forms an individual tumor group (see Müller 1980 for overview).

In 1969, Eck et al. distinguished two groups of undifferentiated cancer types separating primarily small cell undifferentiated carcinomas (oat cell carcinoma) from primarily undifferentiated carcinomas with polymorph cells. These differentiated subdivisions of various histological small cell carcinoma types are also found in the classification of the Veterans Administration Lung Cancer Chemotherapy Study Group (VALG; see Yesner 1973) and the Working Party for Therapy of Lung Cancer (WPL; see Matthews 1973). In these classifications, the generic term of a small cell anaplastic carcinoma of lymphocyte-like type (oat cell type) is separated from the intermediate type (fusiform, polygonal) and others (Israel and Chahinian 1979; Vincent et al. 1977).

The WHO classification of 1967, with fusiform, polygonal, lymphocyte-like, and other cell types, established four subgroups of small cell carcinomas. The present valid WHO classification of 1977 (WHO 1981) includes only three groups:

1) Small cell carcinoma, oat cell type;
2) Small cell carcinoma, intermediate type;
3) Combined small cell carcinoma.

With reference to these presently restrictive proposals for rather precise histological classification of malignant lung tumors, it should be borne in mind that a simultaneous occurrence of histologically very different types of growth within the same tumor is not unusual. For instance, if a small cell carcinoma also shows small proportions of tubular

Table 2. Histological classification of lung tumors (WHO 1981)

I. Epithelial tumors
   A. Benign
      1. Papillomas
         a) Squamous cell papilloma
         b) "Transitional" papilloma
      2. Adenomas
         a) Pleomorphic adenoma ("mixed" tumor)
         b) Monomorphic adenomas
         c) Others
   B. Dysplasia and carcinoma in situ
   C. Malignant
      1. Squamous cell carcinomas (epidermoid carcinoma)
         Variant:
         a) Spindle cell (squamous) carcinoma
      2. Small cell carcinoma
         a) Oat cell carcinoma
         b) Intermediate cell type
         c) Combined oat cell carcinoma
      3. Adenocarcinoma
         a) Acinar adenocarcinoma
         b) Papillary adenocarcinoma
         c) Bronchiolo alveolar carcinoma
         d) Solid carcinoma with mucus formation
      4. Large cell carcinoma
         Variants:
         a) Giant cell carcinoma
         b) Clear cell carcinoma
      5. Adenosquamous carcinoma
      6. Carcinoid tumor
      7. Bronchial gland carcinomas
         a) Adenoid cystic carcinoma
         b) Mucoepidermoid carcinoma
         c) Others
      8. Others
II. Soft tissue tumors
III. Mesothelial tumors
   A. Benign mesothelioma
   B. Malignant mesothelioma
      1. Epithelial
      2. Fibrous (spindle cell)
      3. Biphasic
IV. Miscellaneous tumors
   A. Benign
   B. Malignant
      1. Carcinosarcoma
      2. Pulmonary blastoma
      3. Malignant melanoma
      4. Malignant lymphomas
      5. Others
V. Secondary tumors
VI. Unclassified tumors
VII. Tumor-like lesions
   A. Hamartoma
   B. Lymphoproliferative lesions
   C. Tumorlets
   D. Eosinophilic granuloma
   E. "Sclerosing haemangioma"
   F. Inflammatory pseudotumor
   G. Others

structures or mucus substances and larger tumor cells it is recommended that it is classified according to the predominant cell type (group 2 a in WHO classification of 1977) (Table 2).

If small cell carcinoma proportions as well as distinct squamous cell structures and/or porportions of an adenocarcinoma are present it is recommended to assign this tumor to the group of combined small cell carcinomas (group 2c of WHO classification 1977) (Table 2 and Fig. 5).

For further detail on this subject the reader is referred to the work of Yesner (1973, 1981), Larsson and Zettergren (1976), Sobin (1979), Hirsch et al. (1982), Matthews and Gazdar (1982), and Müller (1984).

Histogenetic Aspects

The great variety of histological types in malignant lung tumors impedes the preparation of a reliable histogenetic concept for individual tumor cells. Starting from the biological concept of disturbed genetic information in tumor cells, a strict classification according to histological criteria seems to be almost preposterous. Malignant tumors show normal structures only in parts deviating from normal in other parts and thus indicating their malignancy. The degree of deviation is different in each tumor; in other words, a histogenetic classification in the strict sense of the word is not possible.

On the other hand, the clinical course of the disease and promising therapy results prove the value of the histological system of classification used at present (Hansen 1980; Li et al. 1981). Despite every objection, morphological studies permit a hypothetical concept of different histogenetic series of tumor development. In the case of squamous cell carcinoma a continuous histogenetic sequence extending from increased basal cell proliferation through squamous metaplasia, dysplastic epithelial changes, and carcinoma in situ up to true cancer can be demonstrated (Harris et al. 1971, 1973).

These preneoplastic lesions, frequently observed in the bronchial system of patients suffering from chronic irritations of the mucosa, may be supported by animal experiment findings (Kuschner and Cashin 1970). In recent years, the cells of the APUD system (amine precursor uptake and decarboxylation system) have gained special importance with regard to histogenetic aspects in connection with the development of SCCL (Pearse 1977). The ultrastructural simultaneous detection of mucous substances, keratin formation, and neurosecretory granules in the same cell, however, makes the derivation of malignant lung tumors of the SCCL type exclusively from APUD cells doubtful (Sidhu 1979). These

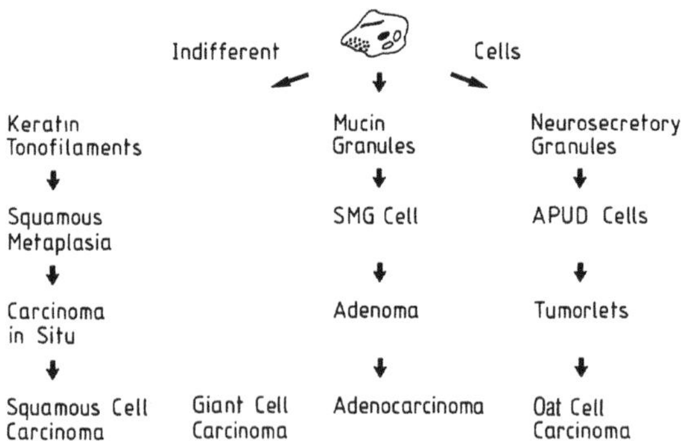

Fig. 4. Hypothetical histogenesis of different types of lung cancer

observations suggest an undifferentiated cell as the possible origin of histologically different malignant lung tumors (Kameya et al. 1980; McDowell and Trump 1981) (Fig. 4).
For further detail the reader is referred to the work of Krompecher (1924), Nasiell (1963a, b), Kracht (1967), Nettesheim and Schreiber (1975), Churg and Warnock (1976), Israel and Chahinian (1979), Ranchod (1977), Said and Mutt (1977), Becci et al. (1978), Hermanek and Gall (1979), Becker et al. (1980), Gazdar et al. (1980), Müller (1981), Kato et al. (1982), Nasiell et al. (1982), Müller and Müller (1983), Reznik-Schüller (1983) and Müller (1984).

Light-Microscopic Findings

"Classic" SCCL consists of small, spindle-shaped or lymphocyte-like tumor cells, poor in cytoplasm and with bare nuclei. The WHO classification of 1977 distinguishes three subtypes with the following characteristics:
1) Small cell carcinoma of oat cell type (lymphocyte-like). The relatively uniform tumor cells of this subtype show cell nuclei with round or oval shapes. Nuclear chromatin is finely distributed, and light microscopy reveals only a thin cytoplasm lining (Fig. 5A). Cells are loosely arranged, the stroma is scarcely developed, and occasionally pseudorosettes appear (Fig. 5B).
2) SCCL of intermediate cell type shows somewhat larger, round, oval, or spindle-shaped and also polygonal nuclei. In this subtype, too, nuclear chromatin distribution is diffuse. Cytoplasm may be more developed. In addition to individually arranged tumor cells, pseudoductal structures can be found (Fig. 5C).
3) The combined small cell carcinoma has small numbers of squamous cells and/or adenocarcinoma structures in addition to the predominating small cell structures (Fig. 5D).

The current opinion is that malignant lung tumors should be classified according to their light microscopic appearance. Additional diagnostic methods, e.g., electron-microscopic studies, are reserved for special questions.

Electron-Microscopic Findings

In many proposals regarding the histological classification, SCCL is described as "undifferentiated" or "anaplastic." The assumption that SCCL is primarily very little differentiated is based on light-microscopic findings. According to electron-microscopic findings, tumor cells of SCCL are capable of a relatively differentiated production (for review see Shimosato et al. 1982).
Electron microscopy has shown that the cells of SCCL include prominent chromatin aggregations in the cell nuclei. The intermediate subtype shows numerous nucleoli. In addition, plentiful polysomes are frequently found and secretory granules are less distinct. These are frequently arranged in pseudopodium-like cytoplasmic offshoots. Occasionally small desmosomes with cell connections between the tumor cells and clearly marked abnormal kinocilia are found. In general, the Golgi's complex is well developed; the rough endoplasmatic reticulum is scarcely distinct in most cases. Some microtubules, tonofibrils, and glandular or rosette-like cytoplasm structures can also be observed (Hattori et al. 1972; Gould and Chejfec 1978) (Fig. 6).

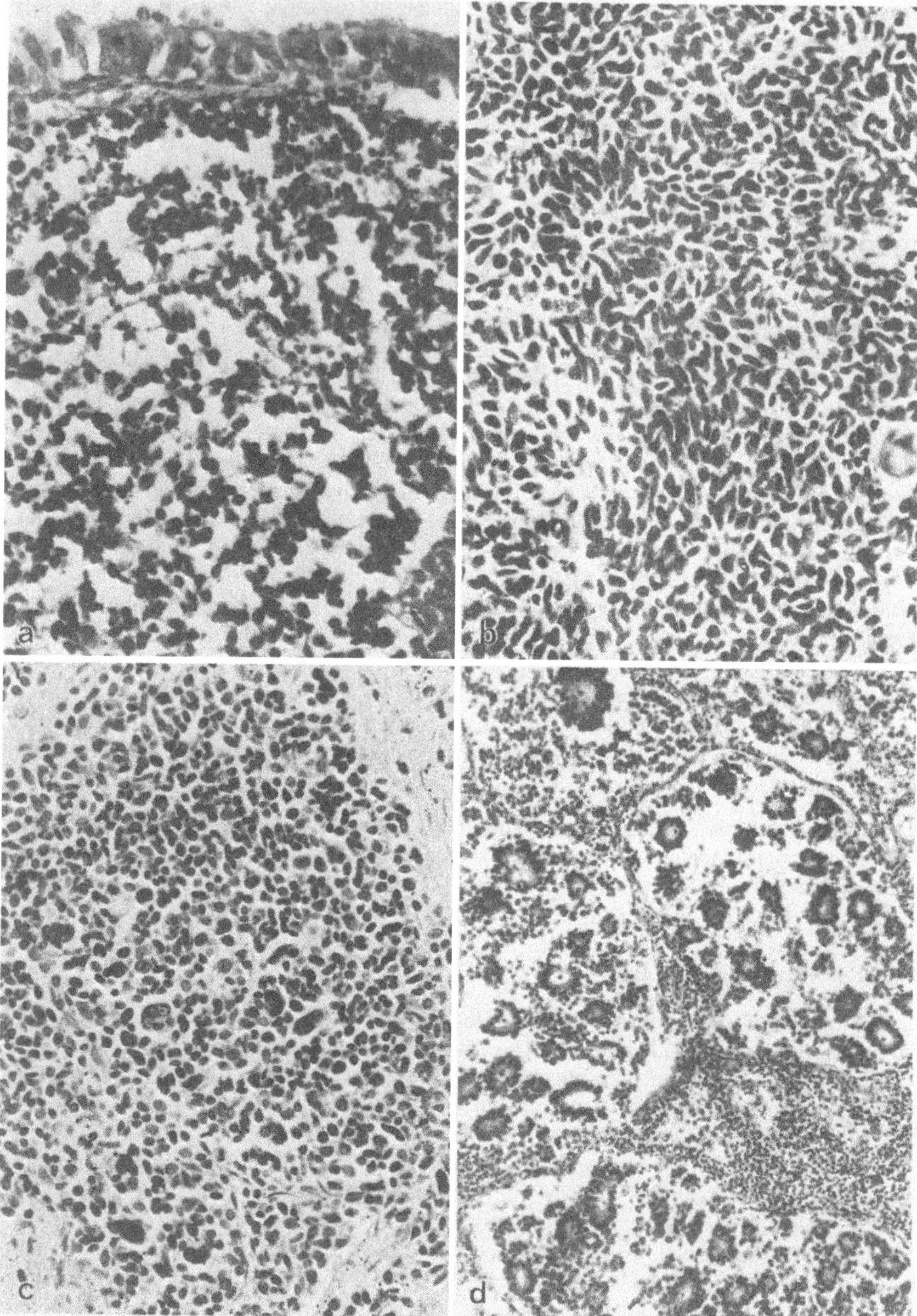

Fig. 5a—d. Histological findings with different types of SCCL. **a** Oat cell carcinoma propagating under epithelial surface; **b** intermediate (fusiform) cell type with pseudo-rosettes; × 350. **c** combined SCCL with giant cells; × 350. **d** combined SCCL with gland-like structures; × 140

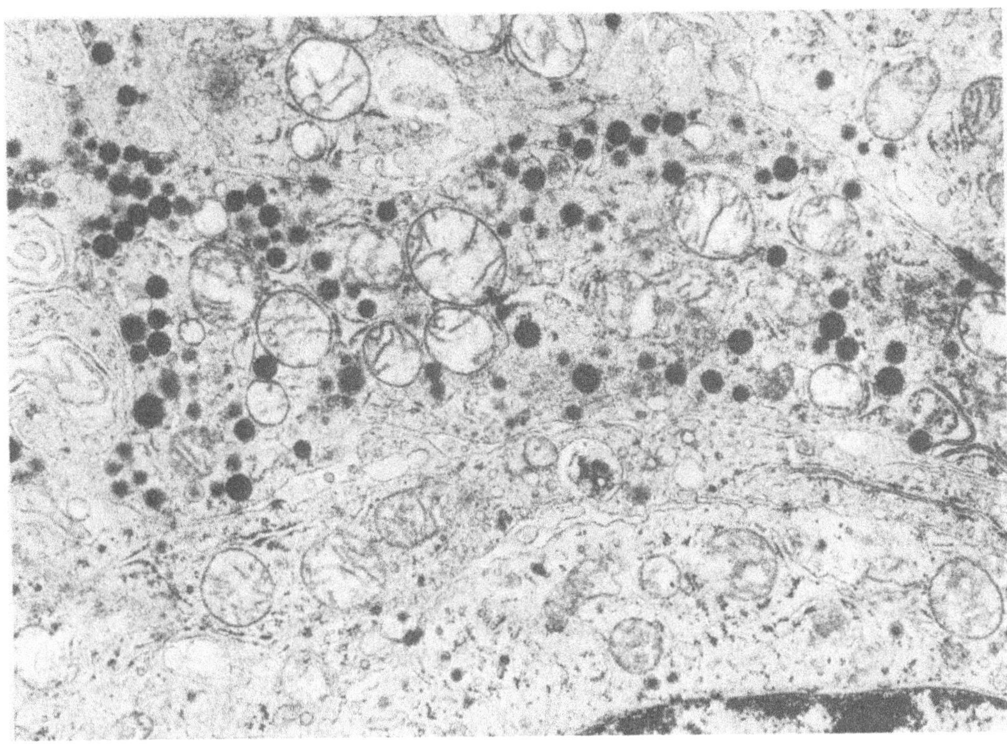

Fig. 6. Submicroscopical illustration of numerous osmiophilic neurosecretory granules in a tumor cell of the APUD system

The frequently demonstrated intracellular secretory granules in SCCL are of particular importance. They are found in 95% of SCCL but their number in different cells varies. Electron-microscopic findings show that they are characteristically localized in pseudopodium-like cytoplasmic offshoots of the tumor cells. These granules are composed of an electron-dense homogeneous core, surrounded by membranes.
These cell products have a diameter of 800–2,000 Å and are surrounded by a membrane with a thickness of 100 Å (McDowell et al., 1976). Recent investigations in SCCL of the oat cell type revealed essentially smaller cytoplasmic granules of 50–190 nm (Shimosato et al. 1982). Thus, these granules are substantially smaller than granules in carcinoid tumors, which have diameters of 90–380 nm. In many cases, the amount of granules can be correlated to clinically observed paraneoplastic syndromes. The ectopic production of ACTH, somatotropin, vasopressin, calcitonin, SMH, and gonadotropins is discussed by Gropp et al. (this volume) (Table 3).
Cells of SCCL show not only neuroendocrine differentiation products but also tonofibrils and desmosomes, the last being typical structures of epithelial differentiation. These electron-microscopic observations match light-microscopic findings of the simultaneous occurrence of different histological features in the same tumor (Fig. 5).
On the other hand, these findings also support suggestions of basal cells of the bronchial mucosa giving rise to SCCL (Feyrter 1954; Bensch et al. 1965, 1968; Gmelich et al. 1967; Terzakis et al. 1972; Tateishi 1973; Höring et al. 1983; Müller 1983).

Table 3. Summary of ultrastructural analysis of small cell carcinoma and carcinoid tumor of the lung

Ultrastructure	Small cell carcinoma	Carcinoid tumor
Nuclear chromatin aggregate	Prominent	Less prominent
Nucleolus	Numerous in intermedate cell type	Less prominent
Polysomes	Abundant	Moderate
Rough endoplasmic membranes	Few	Variable
Golgi complex	Well developed	Well developed
Dense core secretory granule	Few and small	Numerous and large
Microtubules	Numerous in a few cases but usually few	Few
Cytoplasmic pseudopod-like processes	Frequent	Rare
Synaptic vesicles	Rare	Rare
Desmosomes	Occasional	Rare
Intermediate junctions	Frequent	Rare
Small incomplete attachment device	Frequent	Frequent
Tight junctions	Rare except in glandular lumen formation	Rare except in glandular lumen formation
Tonofibrils	More frequent in intermediate cell type	Rare
Glandular (or acinar) structures or true rosettes	Occasional	Occasional
Basement membrane	Often discontinuous	Well defined and continuous
Cilia or basal bodies	Frequent	Rare

Cytological Findings

Cytological detection of SCCL is possible in 70% of cases. Round, oval, and angular cell nuclei are substantial criteria. Nucleus sizes vary between 8 and 12 µm. The chromatin of the tumor cells is reticular, dense, and relatively regular. Nucleoli are scarcely found, and mitoses are detected frequently.

As in histological findings, cytoplasm linings may be confined in relatively few cells. The nucleus-to-plasma ratio is shifted very much in favor of cell nuclei. Mostly, the tumor cells are arranged in loose groups. Artificial cell changes with tadpole-like nuclei and cytoplasmic offshoots are frequent and important cytological criteria. Regressive necrobiotic changes in the nucleus with chromatin aggregation frequently occur (Fig. 7). SCCL of the intermediate cell type also shows somewhat larger spindle-shaped tumor cells.

Sometimes combined SCCL includes cell sizes up to 20 µm. Compared with SCCL, the cells of carcinoid tumors are more uniform in size. Occasionally, in cytological specimens they form rosette-like structures. Cytoplasm is developed to a greater extent and is frequently granulated.

The diagnostic value of cytological investigations in SCCL depends on the site and size of the tumor. Furthermore, the result of the investigation depends heavily on the methods used for collection and preparation of the material.

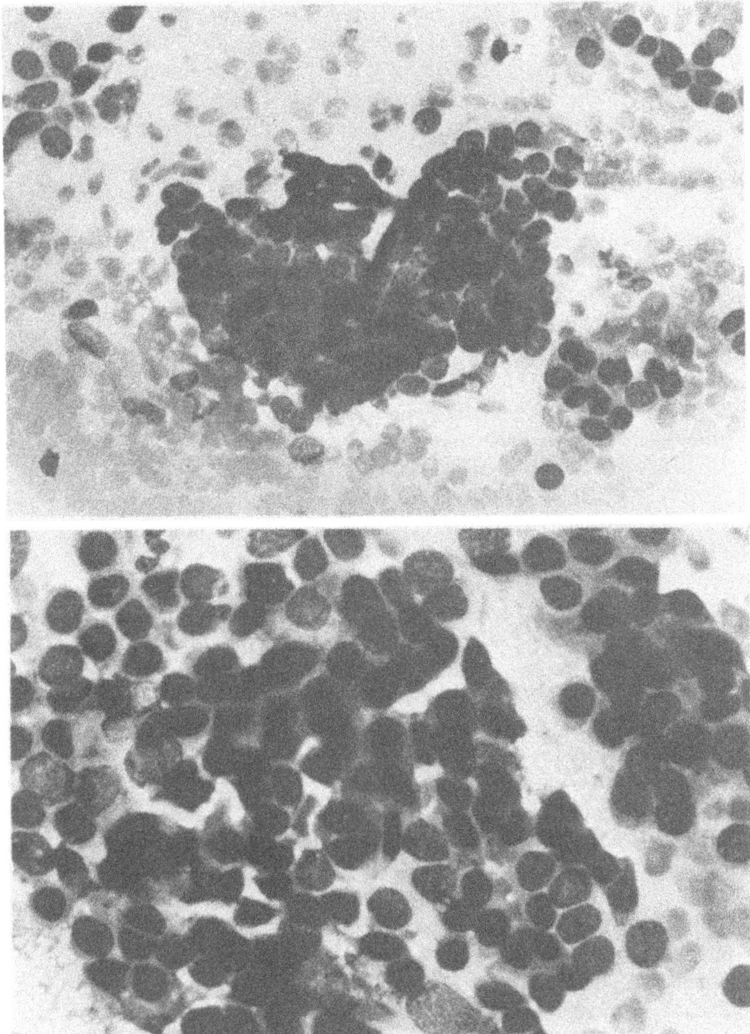

Fig. 7. Cytological findings in SCCL. (*upper panel* × 350; *lower panel* × 560)

For further information on this topic the reader is referred to the work of Koss (1968), Atay and Schlieben (1970), Kahlau (1973), Erozan and Frost (1974), Grunze (1974), Atay (1979), and Takahashi (1981).

References

Atay Z, Schlieben I (1970) Cytologische Diagnostik des Bronchialcarcinoms. Internist (Berlin) 11: 327–343
Atay Z (1979) Cytopathologische Erkennbarkeit früher Neoplasien. Verh Dtsch Krebs Ges 2: 215–232
Barnard WG (1926) The nature of the "oat-celled" sarcoma. J Pathol 29: 241–244

Becci PJ, McDowell EM, Trump BF (1978) The respiratory epithelium IV. Histogenesis of epidermoid metaplasia and carcinoma in situ in the hamster. J Natl Cancer Inst 61: 607–618

Becker KL, Monaghan KG, Silvia OL (1980) Immunocytochemical localization of calcitonin in Kulchitsky cells of human lung. Arch Pathol Lab Med 104: 196–198

Bensch KG, Corrin B, Pariente R, Spencer H, Path FC (1968) Oat cell carcinoma of the lung. Cancer 22: 1163–1172

Bensch KG, Gordon GB, Miller LR (1965) Electron microscopic and biochemical studies on the bronchial carcinoid tumor. Cancer 18: 592–602

Blaha H, Lensch T, Cujnik F (1983) Zur Epidemiologie der endothorakalen Karzinomformen. Kongress der Süddtsch. Ges. für Pneumologie und Tuberkulose, Augsburg, 4 June 1983

Brämer U (1984) Morphometrische Untersuchungen zur Heterogenität bösartiger Lungentumoren. Dissertation, University of Münster

Churg A, Warnock ML (1976) Pulmonary tumorlet. A form of peripheral carcinoid. Cancer 37: 1469–1477

Eck H, Haupt R, Rothe G (1969) Die gut- und bösartigen Lungengeschwülste. In: Uehlinger E (ed) Handbuch der speziellen pathologischen Anatomie und Histologie, vol III/4. Springer, Berlin Heidelberg New York, pp 1–401

Erozan YS, Frost JK (1974) Cytopathologic diagnosis of lung cancer. Semin Oncol 1: 191–198

Fasske E (1970) Histo- und Pathomorphologie der Lungencarcinome. Internist (Berlin) 11: 318–327

Feyrter F (1954) Zur Pathologie des argyrophilen Helle-Zellen-Organes im Bronchialbaum des Menschen. Virchows Arch [Pathol Anat] 325: 723–732

Fischer W (1931) Die Gewächse der Lunge und des Brustfells. In: Hanke F, Lubarsch D (eds) Handbuch der speziellen pathologischen Anatomie und Histologie, Bd. III/3. Atmungswege und Lungen. Springer, Berlin, pp 509–606

Gazdar AF, Carney DN, Rusell EK, Sims HL, Baylin SB, Bunn PA, Guccion JG, Minna JD (1980) Small cell carcinoma of the lung, establishment of continuous clonable cell lines having APUD cell properties. Cancer Res 40: 3502–3512

Giese W (1960) Lungengeschwülste. In: Kaufmann E, Stemmler M (eds) Lehrbuch der speziellen pathologischen Anatomie, Bd. II/3. De Gruyter, Berlin, pp 1904–1944

Gmelich JT, Bensch KG, Liebow AA (1967) Cells of Kultschitzky type in bronchioles and their relation to the origin of peripheral carcinoid tumor. Lab Invest 17: 88–98

Gould VE, Chejfec G (1978) Ultrastructural and biochemical analysis of "undifferentiated" pulmonary carcinomas. Hum Pathol 9: 377–384

Greschuchna D, Maaßen W (1973) Die lymphogenen Absiedlungswege des Bronchialcarcinoms. Thieme, Stuttgart

Grunze H (1974) Zytodiagnostik des Respirationstraktes. In: Soost H-J (ed) Lehrbuch der klinischen Zytodiagnostik. Thieme, Stuttgart, pp 191–275

Hackl H (1969) Über die Metastasen bei 1000 obduzierten Bronchuscarcinomen. Med Monatsschr 23: 490–494

Hansen HH (1980) Management of small cell anaplastic carcinoma. Lung cancer 1980, 2nd world conference, Copenhagen, pp 113–132

Harris CC, Spron MB, Kaufmann DG, Smith JM, Baker MS, Saffiotti U (1971) Acute ultrastructural effects of benzo(a)pyrene and ferric oxide on the hamster tracheobronchial epithelium. Cancer Res 31: 1977–1989

Harris CC, Kaufmann DG, Sporn MB, Saffiotti U (1973) Histogenesis of squamous metaplasia and squamous cell carcinoma of the respiratory epithelium in an animal model. Cancer Chemother Rep 4: 43–54

Hattori S, Matsuda M, Tateishi R, Nishihara H, Horai T (1972) Oat cell carcinoma of the lung. Clinical and morphological studies in relation to its histogenesis. Cancer 30: 1014–1024

Haupt R, Stolper H (1968) Lokalisation und Wuchsform des Bronchialkarzinoms. Zentralbl Allg Pathol 111: 192–201

Hermanek P, Gall FP (1979) Lungentumoren. Kompendium der klinischen Tumorpathologie Bd. 2, G. Witzstrock, Baden-Baden Köln New York

Hirsch FR, Matthews MJ, Yesner R (1982) Histopathologic classification of small cell carcinoma of the lung. Cancer 50: 1360—1366

Höring E, Wörmann B, Büchner T, Müller K-M (1983) Heterogeneity of tumor cells in human bronchial carcinoma. Histological and flow-cytomorphometrical analysis. In: Georgii A (ed) Verhandlung der Deutschen Krebs-Gesellschaft, vol 4, Fischer, Stuttgart, p 846

Hoppe R (1974) Lungenkrebsdiagnostik durch die bronchoskopischen Untersuchungsstellen im Lande Nordrhein-Westfalen (15000 Fälle). Mitteilungsdienst der GBK 2: 3—34

Israel L, Chahinian AP (1979) Lung cancer. Natural history, prognosis and therapy. Academic, London

Kahlau G (1973) Ausgewählte Exfoliativzytologie des Bronchialsystems. Verh Dtsch Ges Pathol 57: 88—95

Kameya T, Tsumuraya M, Adachi I, Abe K, Ichikizaki K, Toya S, Demura R (1980) Ultrastructure, immunohistochemistry and hormone release of pituitary adenomas in relation to prolactin production. Virchows Arch (Pathol Anat) 387: 31—46

Kato H, Konaka C, Hayata Y, Ono J, Iimura I, Matsushima Y, Tahara M, Lei J, Nasiell M, Auer G (1982) Lung cancer histogenesis following in vivo bronchial injections of 20-Methylcholanthrene in dogs. In: Band PR (1982) Early detection and localization of lung tumors in high risk groups. Springer, Berlin Heidelberg New York, pp 69—86

Koss LG (1968) Diagnostic cytology and its histopathological basis. Pitman Medical, London

Kracht J (1967) Pathologie der ektopisch hormonbildenden Tumoren. Verh Dtsch Ges Inn Med 73: 488—495

Kreyberg L (1962) Histological lung cancer types. A morphological and biological correlation. Norwegian University Press, Oslo

Krompecher E (1924) Basalzellen, Metaplasie und Regeneration. Beitr Pathol 72: 163—183

Kuschner M, Cashin S (1970) Pulmonary epithelial tumors and tumor-like proliferation in the rat. In: Nettesheim P, Hanna MG Jr, Deatherage JW Jr (eds) Morphology of experimental respiratory carcinogenesis. USAEC Division of Technical Information, Oak Ridge, pp 203—225 (AEC symposium series 21)

Larsson S, Zettergren L (1976) Histological typing of lung cancer. Application of the World Health Organization classification to 479 cases. Acta Pathol Microbiol Scand [A] 84: 529—537

Li W, Hammar SP, Jolly PC, Hill LD, Anderson RP (1981) Unpredictable course of small cell undifferentiated lung carcinoma. J Thorac Cardiovasc Surg 81: 34—43

Marchesani W (1924) Über den primären Bronchialkrebs. Frankf Z Pathol 30: 158—190

Matthews MJ (1973) Morphologic classification of bronchogenic carcinoma. Cancer Chemother Rep 3: 299—302

Matthews MJ (1979) Problems in morphology and behavior of bronchopulmonary malignant disease. In: Israël L, Chahinian AP (eds) Lung cancer. Natural history, prognosis and therapy. Academic Press, New York, pp 23—62

Matthews MJ, Gazdar AE (1982) Small-cell carcinoma of the lung. Its morphology, behavior and nature. In: Shimosato Y, Melamed MR, Nettesheim P (eds) Morphogenesis of lung cancer, vol II. CRC, Boca Raton, pp 1—14

McDowell E, Trump BF (1981) Pulmonary small cell carcinoma showing tripartite differentiation in individual cells. Hum Pathol 12: 286—294

McDowell E, Barrett LA, Trump BF (1976) Observations on small granule cells in adult human bronchial epithelium and in carcinoid and oat cell tumors. Lab Invest 34: 202—206

Müller K-M (1976) Morphologie und Epidemiologie des Bronchialcarcinoms. Verh Dtsch Krebs Ges 1: 353—378

Müller K-M (1980) Problematik der histologischen Klassifikation des Bronchialcarcinoms. Onkologie 3: 127—132

Müller K-M (1983) Lungentumoren. Pathologie der Lunge II. In: Doerr W, Seifert G, Uehlinger E (eds) Spezielle pathologische Anatomie, vol 16/II, Springer, Berlin Heidelberg New York, pp 1081—1293

Müller K-M (1984) Histological classification and histogenesis of lung cancer. Eur J Respir Dis 65: 4—19

Müller K-M, Müller G (1983) The ultrastructure of preneoplastic changes in the bronchial mucosa. In: Müller K-M (ed) Pulmonary diseases. Springer, Berlin Heidelberg New York, (Current topics in pathology, vol 73)

Nasiell M (1963a) Die Bedeutung der Epithelmetaplasie für die Frage des Bronchialcarcinoms. Ber Dtsch Ges Angew Cytol 1: 40–47

Nasiell M (1963b) The general appearance of the bronchial epithelium in bronchial carcinoma: a histopathological study with some cytological viewpoints. Acta Cytol 7: 97–106

Nasiell M, Carlens E, Auer G, Hayata Y, Kato H, Konaka C, Roger V, Nasiell K, Enstad I (1982) Pathogenesis of bronchial carcinoma, with special reference to morphogenesis and the influence on the bronchial mucosa of 20-methyl-cholanthrene and cigarette smoking. In: Band PR (ed) Early detection and localization of lung tumors in high-risk groups. Springer, Berlin Heidelberg New York (Recent results in cancer research, vol 82)

Nettesheim P, Schreiber H (1975) Advances in experimental lung cancer research. In: Grundmann E (ed) Geschwülste. Springer, Berlin Heidelberg New York, pp 603–691 (Handbuch der allgemeinen Pathologie, vol VI/7)

Pearse AGE (1977) Das diffuse endokrine (parakrine) System: FEYRTER'S Konzept und seine moderne Geschichte. Verh Dtsch Ges Pathol 61: 2–6

Ranchod M (1977) The histogenesis and development of pulmonary tumorlets. Cancer 39: 1135–1145

Reznik-Schüller JM (1983) Spontaneous respiratory tract carcinogenesis. In: Reznik-Schüller HM (ed) Comparative respiratory tract carcinogenesis, vol I. CRC Press, Boca Raton

Said SI, Mutt V (1977) Relationship of spasmogenic and smooth muscle relaxant peptides from normal lung to other vasoactive compounds. Nature 265: 84–86

Schulze W (1974) Geschwülste der Bronchien, Lungen und Pleura. In: Diethelm L, Olson O, Strnad F, Vieten H, Zuppinger A (eds) Handbuch der medizinischen Radiologie, vol IX/4a. Springer, Berlin Heidelberg New York

Shimosato Y, Melamed MR, Nettesheim P (1982) Morphogenesis of lung cancer, vols I, II. CRC, Boca Raton

Sidhu GS (1979) The endodermal origin of digestive and respiratory tract APUD cells. Histopathologic evidence and review of the literature. Am J Pathol 96: 5

Sobin LH (1979) The histological classification of lung tumors. In: Wilkinson PM (ed) Advances in medical oncology, research and education, vol 11. Pergamon, Oxford, pp 5–8

Spencer H (1977) Pathology of the lung, 3rd edn, vol 2. Pergamon, Oxford

Takahashi M (1981) Color atlas of cancer cytology, 2nd edn. Thieme, Stuttgart, pp 267–334

Tateishi R (1973) Distribution of argyrophil cells in adult human lungs. Arch Pathol 96: 198–210

Terzakis JA, Sommers SC, Anderson B (1972) Neurosecretory appearings in cells of human segmental bronchi. Lab Invest 26: 127–132

Vincent RG, Pickren JW, Lane WW, Bross I, Takite H, Houten L, Gutierez AC, Rzepka T (1977) The changing histopathology of lung cancer. A review of 1,682 cases. Cancer 39: 1647–1655

WHO (1981) Histological classification of lung tumors. WHO, Geneva

Yesner R (1973) Observer variability and reliability in lung cancer diagnosis. Cancer Chemother Rep 3: 55–57

Yesner R (1981) The dynamic histopathologic spectrum of lung cancer. Yale J Biol Med 54: 447–456

Possibilities and Limitations of Cytological Diagnoses of Small Cell Bronchogenic Carcinoma

Z. Atay

Medizinische Hochschule Hannover, Pathologisches Institut – Zytologie, Konstanty-Gutschow-Strasse 8, 3000 Hannover 61, Federal Republic of Germany

Introduction

The application of modern therapeutic methods for small cell carcinoma requires precise histomorphological classification prior to treatment. This means, however, that as far as bioptic histology is concerned, certain difficulties are involved; for example, obtaining suitable material from a peripheral location may be problematic. As it is frequently impossible to assess the malignancy or histogenetic type from the generally tiny biopsies taken from small cell carcinoma (Atay 1979; Atay and Preussler 1973; Atay and Schlieben 1969; Barth 1968; Grunze 1966; Herbut and Clerf 1978), the diagnosis often has to be revised at a later stage (Atay and Preussler 1973; Blaha et al. 1965; Graham 1963; Pfaltz and Probst 1953; WHO 1981). Since the 1930s, cytology has assumed ever-increasing significance in the evaluation of exfoliative material and puncture biopsies (Barret 1938/39; Dugeon and Wringley 1935; Farber et al. 1948; Foot 1955; Greschuara et al. 1982; Gupta 1982; Hartmann 1955; Hengstmann 1953). Furthermore, the introduction of imprint preparations means that a precise cytological diagnosis can be made from biopsies too tiny for histological (Atay 1979, 1981a, b; Finsterer and Gahbauer 1972; Fischnaller 1977; Fischnaller et al. 1970; Grunze 1966).

Methods

It is now theoretically possible to obtain cytological material by bioptic methods from all parts of the thoracic cavity. The different kinds of material used are:
1) Purely exfoliative material (sputum and inhalation sputum).
2) Bronchoscopic bioptic material (bronchial secretion, catheter biopsy, bronchoalveolar lavage, bronchial smear, PE smear, perbronchial puncture, transbronchial lung biopsy).
3) Transthoracic material (fine-needle biopsy, imprint preparations from thoracoscopy and mediastinoscopy, intraoperative imprint preparations obtained during thoracotomy)

The selection of the method to be used depends largely on the site, extent, and nature of the tumor. Although the small cell carcinoma has a predominantly central localization, it grows extremely quickly and necroses tend to develop. Since the neoplastic cells show minimal cohesion with one another, cells are soon discovered in exfoliative material and very early metastases are seen, particularly in the mediastinal lymph nodes. Good diagnostic results can therefore be obtained from examination of exfoliative material.

However, bronchoscopy is preferable because it allows a visual check, which makes it possible to assess operability, take perbronchial punctures and, most important, obtain biopsies for histology and imprint cytology. Moreover, bronchoscopy makes a greater amount of tissue available and takes into account any intramural growth of the small cell carcinoma. At present, cytological evaluation of PE smears and perbronchial punctures is unfortunately limited to a few centers only (Atay 1979, 1981a, b; Atay and Brandt 1975; Finsterer and Gahbauer 1972; Fischnaller 1977; Lopez-Cardozo 1954; Preussler and Pfannkuch 1979; Roglić 1976; Salzer 1970). Sputum, bronchial secretion, catheter biopsies, and brush smears are only useful if the tumor is located peripherally.

Staining Techniques

Methods of staining include those of Pappenheim and Papanicolaou. Use of the Pappenheim stain is recommended for puncture material and imprint preparations. Only by this method can a clear color grading of the cytoplasm and nucleoli be obtained, which is indispensable for differentiation of small cell carcinoma from lymphoma (Atay 1979, 1981a, b; Fischnaller 1977; Grunze 1955; Kanhouwa and Matthews 1976; Lopez-Cardoza et al. 1967). This stain also enables malignancy to be determined, even when only a few cells are available. On the other hand, the Papanicolaou stain is considerably more suitable for finding tumor cells in exfoliative material, because of its increased transparency. It is also better for revealing nuclear hyperchromatism. Only rarely are special stains necessary (silver impregnation, immunoperoxidase), particularly for distinguishing small cell carcinoma from apudoma (Müller-Hermelink and Fritsch 1979), but also from Ewing sarcoma (PAS reaction).

Cytomorphology

The first authors to lay down cytological criteria for small cell bronchogenic carcinoma were Dudgeon and Wrigley in 1935. Since then, this tumor type has been described in great detail (Dröse and Bayer 1976; Farber et al. 1948; Finsterer and Gahbauer 1972; Foot 1955; Gagneten et al. 1976; Gowar 1943; Grunze 1968; Gupta 1982; Kluge 1965; Lopez-Cardozo 1954, 1975; Marrommatis 1965. The varying tissue and cellular differentiation of the small cell carcinoma, which Barnard (1926) noted quite early, enabled the WHO (World Health Organization; 1981) to establish three subgroups:
1) Oat cell carcinoma.
2) Small cell carcinoma of intermediary cell type.
3) Combined oat cell carcinoma.

This organization stated that although the small cell carcinoma could be recognized by cytology, a more precise classification of the tumor into subgroups could only be accomplished by histological means. Some cytopathologists, however, are of the opinion that an exact subdivision is also possible by cytological methods (Atay 1979, 1981b; Lopez-Cardozo 1975; Pfitzer et al. 1978). Since 1967 we have been recording our findings according to the WHO classification, and similar uses of these subdivisions are also found, if only occasionally, in the literature (Suprun et al. 1980; Zaharopoulos et al. 1982). Figure 1 shows some of the cellular criteria for small cell carcinoma of the *oat cell type*. The tumor cells reveal a maximally shifted nucleus-to-plasma ratio in favor of the nucleus. Slight anisokaryosis and irregular and angular nuclear borders are frequently seen. The

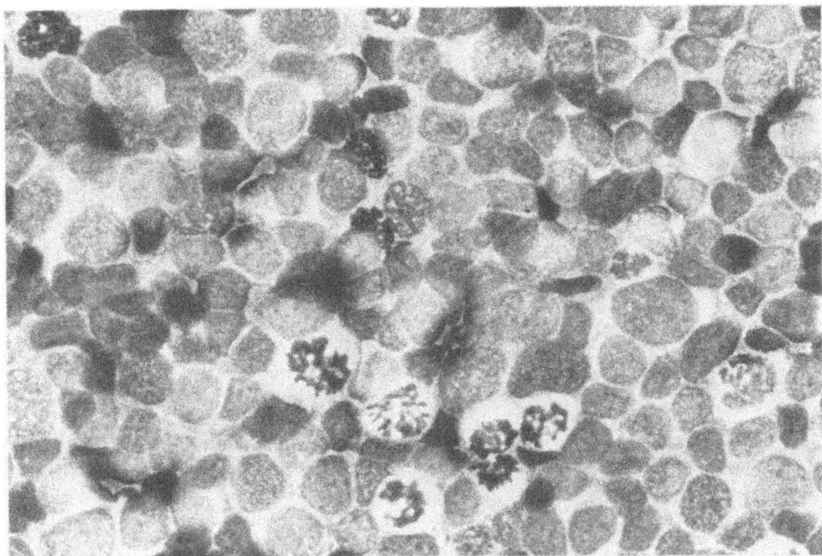

Fig. 1. Small cell carcinoma of oat cell types, examined in perbronchial punctate. Large quantities of small cells with maximum shift of nucleus-plasma ratio; irregular, angular nuclear borders and finely reticulated, dense chromatin structure. Abundant mitoses, no organoid structures. Pappenheim, × 160 (Zeiss)

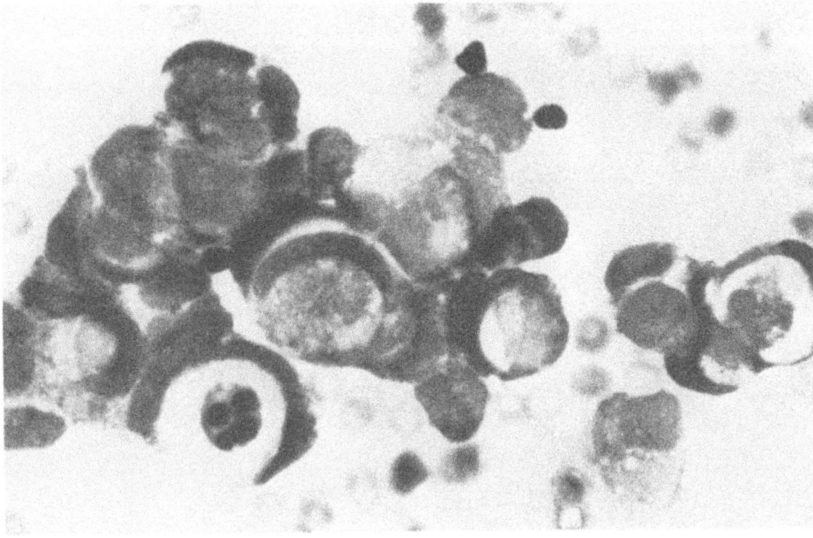

Fig. 2. Small cell carcinoma of oat cell type, examined in pleural punctate. Small to medium-sized tumor cells with extreme nuclear hyperchromatism and reticular chromatin structure; frequent cell cannibalism and bud-shaped orientation. Pappenheim, × 200 (Zeiss)

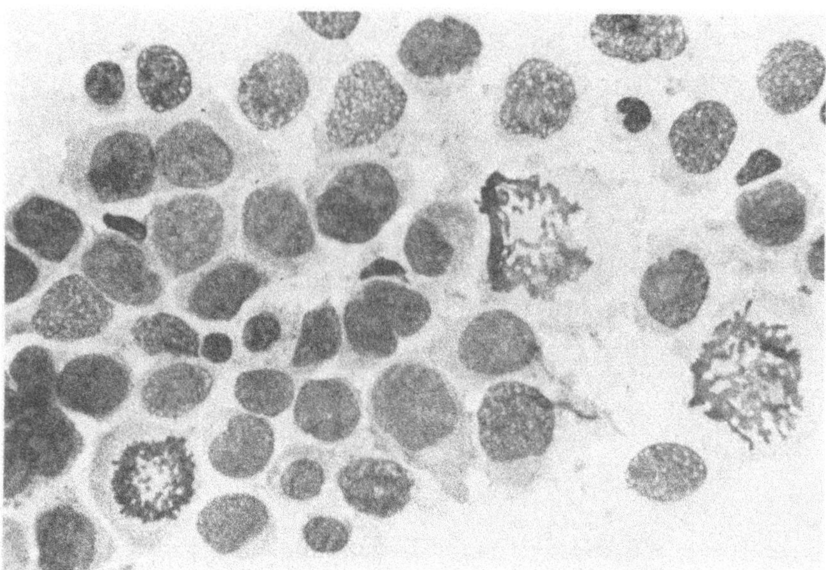

Fig. 3. Small cell carcinoma of intermediary cell types, examined in PE smear from bronchus. Separate medium-sized tumor cells with sparse, poorly defined cytoplasm showing light-blue staining and finely reticulated chromatin structure; some mitoses. Pappenheim, × 200 (Zeiss)

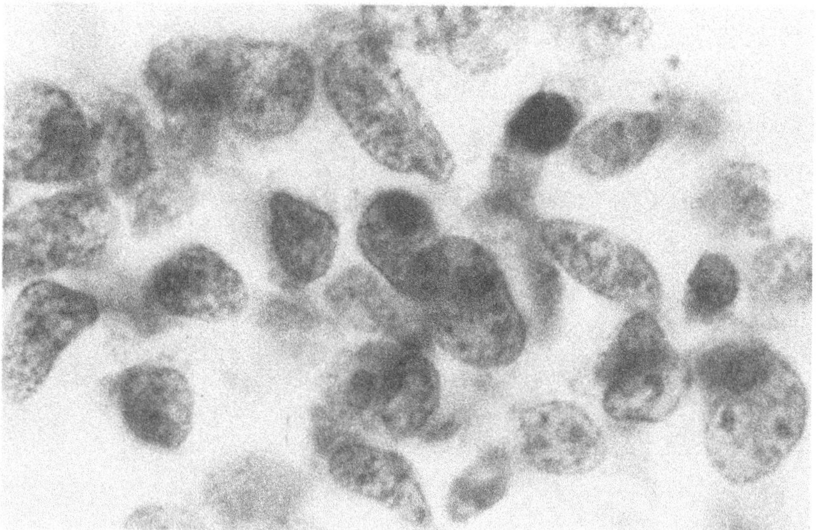

Fig. 4. Small cell carcinoma of intermediary cell type with predominance of spindle cells, examined in bronchial secretion. Predominance of thick, spindle-like tumor cells with irregular nuclear borders and granular, essentially regular chromatin distribution. Almost all tumor cells have bare nuclei. Papanicolaou, × 250 (Zeiss)

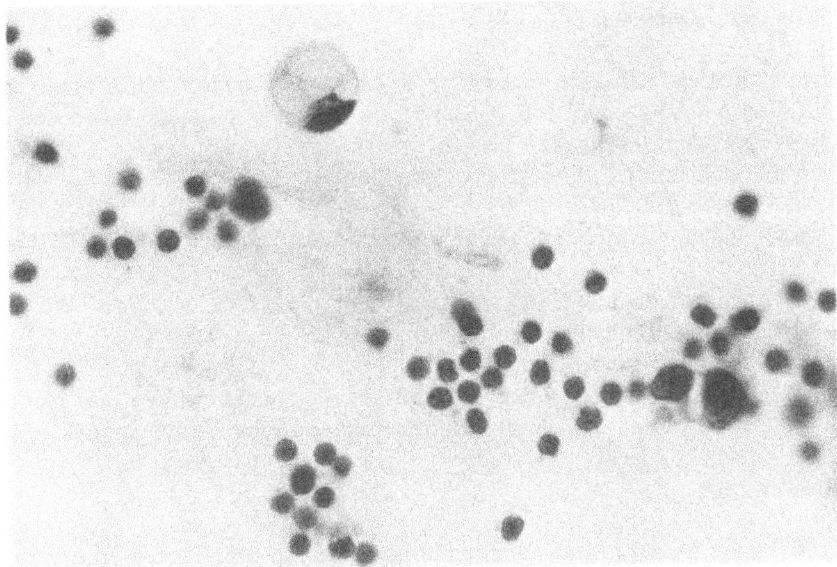

Fig. 5. Combined oat cell carcinoma (oat cell carcinoma plus adenocarcinoma), examined in the sputum. Abundant small tumor cells with maximum shift in nucleus-to-plasma ratio and extreme nuclear hyperchromatism (lymphocyte-like). Three large tumor cells with irregular nuclear borders, eccentric nuclear orientation and vacuolized cytoplasm. Papanicolaou, × 80 (Zeiss)

chromatin structure shows considerable reticulation. The light blue nucleoli are not enlarged and, indeed, tend to be small. The light blue to colorless cytoplasm is so easily damaged that the tumor cells generally show no cytoplasmic demarcations and appear to have no cytoplasm. As the nuclei may also be injured easily, increased cell particles are found in the smear and cells are rarely intact. The tumor cells are usually isolated but may form groups or organoid structures. A whorled or trough-shaped arrangement is particularly characteristic of effusions (Fig. 2).

The characteristics of small cell carcinoma of the *intermediary cell type* are shown in Fig. 3. The cell picture is less homogeneous; large cells predominate with polygonal cytoplasm in moderate to abundant amounts. The nucleus is often centrally situated and shows reticulated chromatin. An abundance of spindle-shaped cells (Fig. 4) may develop side by side with smaller tumor cells, the so-called oat cells. The other criteria are the same as for oat cell carcinoma. In both types, the tumor cells may occasionally form rosette patterns.

In *combined oat cell carcinoma*, as well as classic small cell tumor cells, cells with squamous (Fig. 5) or adenomatous differentiation are also seen. Thus, keratinized and also nonkeratinized squamous cells, mucus-producing cells and cell groups showing papillary or acinar patterns may be seen side by side.

The cytomorphological criteria for the primary small cell bronchogenic carcinoma are largely identical with those for its metastases.

Differential Diagnosis

If small atypical cells are seen in the smear, the following cell or tumor groups must be excluded before a small cell bronchogenic carcinoma can be diagnosed:
1) Hyperplastic basal cells.
2) Small cell tumors of varying origin
   a) Carcinoid;
   b) Malignant lymphoma (lymphoblastic, centroblastic, small cell histiocytic sarcoma);
   c) Ewing's sarcoma;
   d) Neuroblastoma, medulloblastoma;
   e) Round cell fibrosarcoma.
3) Tumor which, although not originally of the small cell type, has metastases showing portions of small cell differentiation (adenocarcinoma, squamous cell carcinoma).

Results

Optimum preparation and staining methods and the experience of the cytopathologist and the assistants all play a considerable role in the detection of neoplastic cells. Besides these factors, however, the diagnosis is also largely dependent on the type of material available and the possibilities and limitations of cytomorphology itself, i.e., to what extent the cytological criteria allow recognition of the malignancy and histogenesis of small cell bronchogenic tumors.

Besides purely exfoliative substances, specific, productive biopsy materials, such as punctures and imprint cytology, are available for examination. For an assessment of the capabilities and limitations of cytomorphology, a swab from the biopsy may be examined and the results compared with those of histological examination of the same biopsy. Our group was able to detect neoplastic cells in 186 swab preparations from 188 cases (99%) in which tumor material had been seen morphologically (Atay and Preussler 1973). Since we could not diagnose malignancy in the other two cases even at a repeat examination, the smears had evidently been prepared from portions of benign tissue. Similarly, when we examined smears from 1,579 cases declared benign after histological tests we did not make a single diagnosis of malignancy. I was also able to define correctly the histogenesis of 235 (90.7%) of 259 small cell carcinoma cases (Atay 1981a), a result which could be improved upon with increasing experience. The type of material used for determining the histogenesis of small cell bronchogenic carcinoma does not play a significant role since, in contrast to other tumors (squamous cell carcinoma, adenocarcinoma), the structure and morphology of the neoplastic cells are largely identical in all areas of the tumor. On the other hand, as already mentioned, the experience of the cytopathologist and the assistants is of great importance. The clinician, however, is essentially interested in whether a carcinoma is of the small cell type or not. I was able to identify correctly 181 (87.5%) of 204 tumors diagnosed histologically as small cell carcinoma. Of 891 tumors not of this variety, I erroneously diagnosed only 5 (0.56%) as small cell carcinoma. This indicates that when a cytologist makes a diagnosis of "non-small-cell carcinoma" he agrees in 99.4% of cases with the histological findings, while 12.5% of his diagnoses of "small cell carcinoma" are false. False diagnoses were made for 26 small cell tumors, 18 of which were mistaken for squamous cell carcinoma and 4 for adenocarcinoma or large cell carcinoma. Five cases were diagnosed by cytology as small cell carcinoma, while they were found by histology to be squamous cell carcinoma.

Table 1. Accuracy rates for aspirates and catheter biopsies[a] according to the various WHO stages of operability

Stage of operability	n	Positive cytology		
		Aspirate	Catheter biopsy	Combined
1 Operability technically and prognostically very good	4	–	1 (25%)	1 (25%)
2 Operable both technically and prognostically	7	2 (29%)	5 (71%)	5 (71%)
3 Operable technically, inoperable prognostically	10	2 (20%)	4 (40%)	4 (40%)
4 Inoperable both technically and prognostically	24	9 (38%)	14 (58%)	16 (67%)
	45	13 (29%)	24 (53%)	26 (58%)

[a] Cytological examination of catheter biopsies is superior to that of aspirates

Dependence of Malignancy Determination on Type of Material

For precise determination of the rate of accurate diagnosis for exfoliative substances, all examination material must be taken on the same occasion so that conditions are uniform. Table 1 compares the findings recorded in examinations of bronchial secretion and in catheter biopsies according to the WHO tumor stage classification. All these small cell tumors were located peripherally. A clear correlation is seen between tumor stage and detection rate: the more advanced the tumor is, the more likely it is to be identified. On the whole, the results of catheter biopsy examinations are about 24% better than those of aspirates. The accuracy rate increases to 58% when a combination of both methods is used. It is extremely difficult to compare our findings with others in the literature, because of the highly variable nature of each individual tumor (size, location, operability), and for this reason no reliable rates of accuracy can be given. The efficiency of cytology in the examination of exfoliative material is shown in Table 2. Of 11 small cell bronchogenic tumors, Bedrossian and Rybka (1976) identified 45.5% from sputum examination, 73% from brush biopsies, and 64% from aspirates. Holbro et al. (1978) diagnosed 56% of 52 cases from the sputum, 29% from brush biopsies, and 69% from aspirates. Of the diagnoses of Pfitzer et al. (1978), 42% (of 38) of those made from sputum examinations and 15% (of 33) from aspirates were positive. The considerable quantitative differences seen here may be explained by the variations in the methods of obtaining material. For example, Holbro's brush biopsies, unlike Bedrossian's, were nonspecific; Pfitzer always made several sputum examinations and only included operable cases. Bedrossian's results probably appear more favorable than our own because he does not separate central and peripheral tumors. As already mentioned, the type of material plays little part in the determination of histogenetic tumor type. As becomes clear in Table 3, accuracy rates vary between 70% and 97%, the median being about 90%.
To evaluate the reliability of cytological methods, the results of histological determinations of malignancy and histogenesis of small cell bronchogenic carcinoma are compared with those obtained by applying imprint cytology to the same material.

Table 2. Cytological accuracy rates[a] for determination of malignancy in small cell bronchogenic carcinoma by examination of exfoliative material

Authors/years	Sputum		Brush		
	+/n	%	+/n	%	
Bedrossian and Rybka (1976)	5/11	45.5	8/11	73	Peripheral and central tumors, specific brush material
Holbro et al. (1978)	29/52	56	15/52	29	Peripheral and central tumors, nonspecific brush material
Pfitzer et al. (1978)	16/38	42	–	–	Only operable tumors
Own group (unpublished)	–	–	–	–	Only peripheral tumors

Authors/years	Aspirate		Catheter biopsy		Remarks
	+/n	%	+/n	%	
Bedrossian and Rybka (1976)	7/11	64	–	–	Peripheral and central tumors, specific brush material
Holbro et al. (1978)	36/52	69	–	–	Peripheral and central tumors, nonspecific brush material
Pfitzer et al. (1978)	5/33	15	–	–	Only operable tumors
Own group (unpublished)	13/45	29	29/45	53	Only peripheral tumors

[a] The rate of accuracy is dependent on the nature of the tumor and the method of obtaining material

Of Diagnosis of Small Cell Bronchogenic Carcinoma by Probability Histological Examination of Neoplastic Tissue. Experience (Atay et al. 1981; Kluge 1965) has shwon that the smaller the biopsy, the more difficult it is to make a histological diagnosis; this is particularly true of peripheral tumors (Graham 1963). As we have demonstrated in our earlier work (Atay 1979; Atay and Preussler 1973; Atay et al. 1981), catheter biopsy material should not be prepared for histology when little material is available. This is true, for example, of inflammatory processes where the small tumor cells can hardly be distinguished from the inflammatory cells. A comparison of histology and imprint cytology reveals that in the determination of malignancy, the former has a rate of accuracy of 80% and the latter of 90% (Table 4). For other epithelial tumors the accuracy rate of histology is 10% better, at 90%.

Histological Determination of Histogenetic Tumor Type. According to the literature, histological determination of histogenesis in small biopsies from the thoracic cavity has a fairly high rate of error, quite apart from differences in interpretation (Salzer 1970). Brandt

Table 3. Review of accuracy rates for the determination of histogenesis of small cell bronchogenic carcinoma by cytology

Authors/year	Cases n	Correct diagnosis of type		Remarks
		n	$\%$	
1 McBurney et al. (1951	14	10	70	Sputum and aspirates
2 Fischnaller et al. (1970)	18	14	78	Diverse material
3 Finsterer and Gahbauer (1972)	37	35	95	Lung punctures
4 Atay and Brandt (1975)	61	56	92	Aspirates
5 Dröse and Bayer (1976)	14	12	86	Sputum and aspirates
6 Kanhouwa and Matthews (1976)	19	17	90	Diverse material
7 Suprun et al. (1980)	35	34	97	Diverse material
8 Atay (1981)	259	235	91	Imprint preparations from thoracic cavity
9 Atay et al. (1981)	20	19	95	Catheter biopsies
10 Gupta (1982)	76	70	92	Sputum

Table 4. Results of combined histological and cytological examinations[a] of biopsies from small cell bronchogenic carcinoma

Type of examination	Determination of malignancy	
	n	$\%$
Histology	150	80
Cytology	186	99
Positive histology and cytology	148	78
Positive histology only	2	1
Positive cytology only	38	21
Positive cytology or histology	188	100

[a] The histological results were improved by 20%

and Loddenkemper (1981), therefore, fittingly used the term "random sample." To assess the reliability and usefulness of preoperative histology and cytology, their results must be comparised with those of operative histology. Greschuchna et al. (1983) showed that 72% of preoperative histological diagnoses agree with the findings operative histological tests. Similar results can be found elsewhere in the literature. We found that 95.8% of our own cytological diagnoses were identical with the findings of operative histology of small cell bronchogenic carcinoma, while the accuracy rate for bioptic histology was only 87.5%. This means that for the determination of histogenesis, the histological examination of biopsy material is surpassed by cytology. Such success may be further increased by using cytochemical examinations and electron microscopy.

Summary

We investigate the possibilities and limitations of cytological examinations of small cell bronchogenic carcinoma by considering the results of cytodiagnoses of malignancy and histogenesis from both exfoliative and biopsy material. When a diagnosis is made from biopsy material the accuracy rate is found to be 99% for cytology, indicating that the rate for histology could be improved by as much as 20%. In peripheral small cell bronchogenic carcinoma, examination of catheter biopsies was seen to be superior to that of sputum or aspirates. When exfoliative material was used, however, only about half the cases admitted of diagnosis. Determination of histogenesis appears to depend largely on the type of material examined, and the accuracy rates vary between 70% and 95%. It is clear that the results obtained with cytological methods are considerably better than those obtained with preoperative and intraoperative histological methods.

References

Atay Z (1979) Cytopathologische Erkennbarkeit früher Neoplasien. In: Georgii A (ed) Frühe Tumoren in Diagnostik und Therapie. Verh Dtsch Krebs Ges, vol 2, Fischer, Stuttgart, pp 215–231

Atay Z (1981a) The reliability of cytodiagnosis in determining malignancy and histogenetic tumor type. In: Nakhosteen JA, Maassen W (eds) Bronchology: research, diagnostic and therapeutic aspects. Martinus Nijhoff, The Hague, pp 37–42

Atay Z (1981b) Zytologie des Bronchialsekretes. In: Informationen aus Lehre und Forschung 1/1981. Freie Universität, Berlin, pp 165–194

Atay Z, Brandt HJ (1975) Ergebnisse cytologischer Untersuchungen des Bronchialsekretes bei Lungentumoren im Verhältnis zum Tumorstadium. Dtsch Med Wochenschr 100: 1269–1274

Atay Z, Freise G (1971) Die Treffsicherheit der intraoperativen Cytodiagnostik bei Thoraxoperationen. Pneumologie 145: 423–427

Atay Z, Gräfin von Schlieben I (1969) Cytologie intrathorakaler Organe. Med Lab 22: 271–283

Atay Z, Preussler H (1973) Ergebnisse der vergleichenden Zytologie und Histologie von 921 malignen Tumoren aus 2500 Biopsien im Thorax. Verh Dtsch Ges Pathol 57: 360–362

Atay Z, Hammesfahr R, Worch R, Ekinci C (1981) Cytologische Diagnose der peripheren Lungentumoren durch kombinierte Auswertung der Katheterbiopsien und der gezielten Absaugungen. Prax Pneumol 35: 220–224

Barnard WG (1926) The nature of the "oat-celled sarcoma." J Pathol Bacteriol 29: 241–244

Barret NR (1938/39) Examination of the sputum for malignant cells and particles of malignant growth. J Thorac Surg 8: 169–183

Barth L (1968) Die endoskopische Diagnose des Bronchialkarzinoms. Ergebnisse von 9841 diagnostischen Bronchoskopien bei 2767 histologisch gesicherten Bronchialkrebsen. Arch Geschwulstforsch 32: 81–94

Bedrossian C, Rybka DL (1976) Bronchial brushing during fiberoptic bronchoscopy for the cytodiagnosis of lung cancer. Comparison with sputum and bronchial washings. Acta Cytol 20: 446–453

Blaha H, Kahlau G, Ungeheuer E (1965) Kleinzellige Bronchialkarzinome. Thieme, Stuttgart

Brandt HJ, Loddenkemper R (1981) Voraussetzungen für die operative, radiologische und zytologische Behandlung intrathorakaler Tumoren. Prax Pneumol 35: 851–864

Dröse M, Bayer E (1976) Zytologische Klassifizierung von Bronchialkarzinomen. Dtsch Med Wochenschr 39: 1417–1420

Dröse M, Bayer E, Präuser H (1978) Sputum und Bronchialsekretzytologie. Dtsch Med Wochenschr 103: 244–248

Dugeon LS, Wrigley LH (1935) On the demonstration of particles of malignant growth in sputum by means of the wet film method. J Laryngol Otol 50: 752−763

Farber M, Martimer S, Benioff A, Frost JK, Rosenthal M, Tobias G (1948) Cytology studies of sputum and bronchial secretions in primary carcinoma of the lung. Dis Chest 14: 633−664

Finsterer H, Gahbauer H (1972) Zur Differentialdiagnose verschiedener Tumorformen in der Lungenzytologie. Beitr Pathol 145: 350−364

Fischnaller M (1977) Cytological fine diagnosis in pulmonology. Österreichische Zeitschrift für Onkologie 4: 65−80

Fischnaller M, Hack H, Schwarzenberg E (1970) Die Aussagewerte der Cytologie in der Lungenpathologie. Prax Pneumol 24: 145−157

Foot CN (1955) Cytologic diagnosis in suspected pulmonary cancer. Am J Clin Pathol 25: 223−241

Gagneten CB, Geller CE, Seanz MC (1976) Diagnosis of bronchiogenic carcinoma. Acta Cytol 20: 530−536

Gowar FJ (1943) Carcinoma of the lung. Br J Surg 39: 193−200

Graham RM (1963) The cytologic diagnosis of cancer, 2nd edn. Saunders, Philadelphia London

Greschuchna D, Kasparek R, Kappes R (1983) Ein Vergleich der Histologien präoperativer Bronchusbiopsien und Mediastinalbiopsien mit Lungenresektaten von Bronchialkarzinomen. Prax Klin Pneumol 37: 862−865

Grunze H (1955) Klinische Cytologie der Thoraxkrankheiten. Enke, Stuttgart

Grunze H (1966) Derzeitiger Stand der Zytodiagnostik bei Erkrankungen des Thorax. Dtsch Med Wochenschr 91: 1476−1483

Grunze H (1968) Methoden und Indikationen bioptischer Untersuchungen bei Verdacht auf Lungenkrebs. GBK-Mitteilungen 5: 82−95

Gupta RK (1982) Value of sputum cytology in the diagnosis and typing of bronchogenic carcinoms, excluding adenocarcinomas. Acta Cytol 26: 645−648

Hartmann P (1955) Die Cytologie des Bronchialsekretes. Thieme, Stuttgart

Hengstmann H (1953) Die Cytodiagnose des Bronchialcarcinoms mit Hilfe der gezielten Bronchialabsaugung. Klin Med 150: 283−312

Herbut PA, Clerf LH (1946) Bronchiogenic carcinoma diagnosis by cytologic study of bronchioscopically removed secretions. JAMA 130: 1006−1012

Holbro P, Dalquen P, Perruchoud A, Herzog H (1978) Zytologische und histologische Untersuchungsmethoden bei Lungentumoren. Dtsch Med Wochenschr 103: 17−20

Jäger J (1978) The diagnosis of the main histological types of the lung cancer. Arch Geschwulstforsch 48: 240−244

Kanhouwa SB, Matthews MJ (1976) Reliability of cytologic typing of lung cancer. Acta Cytol 20: 229−232

Kauder H (1974) Probleme und Schwierigkeiten bei der Zytodiagnostik des Bronchialkarzinoms. Z Erkr Atmungsorgane 138: 161−171

Kluge J (1965) Die histologische Beurteilung des Katheterbiopsie-Aspirates beim Bronchialkarzinom. Z Tuberkulose 124: 97−100

Langer C, Lobenwein E, Dokulil E (1964) Die Bedeutung des zytologischen Befundes für die Frühdiagnose des Lungenkrebses. Prax Pneumol 18: 688−690

Lopez-Cardozo P (1954) Clinical cytology. Stafleu, Leyden

Lopez-Cardozo P (1975) Atlas of clinical cytology. Targa, s'Hertogenbusch

Lopez-Cardozo P, de Graaf S, de Boer MJ, Doesburg N, Kapsenberg PD (1967) The results of cytology in 1,000 patients with pulmonary malignancy. Acta Cytol 11: 120−131

Mavrommatis F (1965) Zur Differentialdiagnose der Tumortypen des Bronchialkarzinoms im zytologischen Ausstrich nach Papanicolaou. Thoraxchirurgie 12: 412−421

McBurney RB, Robert MD, McDonald MD, John R, Clagett MD (1951) Bronchiogenic small cell carcinoma. J Thorac Surg 22: 62−73

Müller-Hermelink HK, Fritsch H (1979) Histochemische und Immunologische Befunde an APUD-Zellen im menschlichen Bronchialsystem. In: Morphologie und Pathomorphologie der Lunge. Festschrift zum Festcolloquium. Freie Universität Berlin, pp 115−126

Pfaltz CR, Probst R (1953) Die normale und pathologische Cytologie des Bronchialsekretes und der Oesophagusschleimhaut. Arch Klin Exp Ohren-Nasen-Kehlkopfheilkd 164: 225–296

Pfitzer P, Sygula E, Bernhardt-Huth D (1978) Beiträge zu den morphologischen Grundlagen der Zytologie des Respirationstraktes. Arch Geschwulstforsch 48(3): 220–232

Preussler H, Pfannkuch F (1979) Zur Problematik der Klassifikation der Lungentumoren. In: Morphologie und Pathomorphologie der Lunge, Festschrift zum Festcolloquium. Freie Universität Berlin, pp 195–208

Roglić M (1976) Cytological examination of forceps biopsy materials obtained at bronchoscopy. 6th congress of the European Federation of Cytologic Societies, Weimar

Roglić M (1981) Forceps-biopsy material. A sample for histologic and cytologic examination. In: Nakhosteen JA, Maassen W (eds) Bronchology: Research, diagnostic and therapeutic aspects. Martinus Nijhoff, The Hague, pp 37–42

Salzer G (1970) Klinische Überlegungen zur Histologie des Bronchialcarcinoms. Verh Ber Dtsch Tg 24: 131–139

Sigismund G, Atay Z, Löblich HJ (1978) The diagnostic value of combined bronchus biopsy and contact smear cytology of the biopsy specimen in lung cancer. Cancer Cytol 18: 9–12

Strupler W (1955) Die Cytologie des Tracheo-Bronchialsekretes. Fortschr HNO-Heilk 3: 280–287

Suprun H, Pedio G, Ruttner JR (1980) The diagnostic reliability of cytologic typing in primary lung cancer with a review of the literature. Acta Cytol 24: 494–500

WHO (1981) Histological typing of lung tumours. WHO, Geneva

Worch J-K, Worch R, Atay Z (1983) Zytologische Bestimmung des histogenetischen Typs der Lungentumoren. Prax Klin Pneumol 37: 839–841

Zaharopoulos P, Wong JY, Stewart GD (1982) Cytomorphology of the variants of small cell carcinoma of the lung. Acta Cytol 26: 800–808

Cytogenetics of Human Small Cell Lung Cancer

J. Whang-Peng and E. C. Lee

Cytogenetic Oncology Section, MB, COP, DCT, NCI, National Institutes of Health, Bethesda, MD 20205, USA

Introduction

The combination of chemo- and radiotherapy has resulted in prolonged survival and potential cures in patients with some neoplastic diseases. Small cell lung cancer (SCLC) is one of those neoplasms in which over 90% of the patients respond favorably to treatment, about 10% being cured. However, patients with non-small-cell lung cancer (non-SCLC) have a much lower response rate to current therapeutic regimens. Therefore, accurate diagnosis is extremely important. With the advent of chromosome banding techniques in the early 1970s (Caspersson et al. 1968; Seabright 1971), it became possible to identify chromosomal abnormalities that were specific for certain neoplastic diseases, including the Philadelphia or Ph[1] chromosome in chronic myelogenous leukemia (CML) (Nowell and Hungerford 1960), t(15;17) in acute progranulocytic leukemia (APL) (Rowley et al. 1977), t(8;14) in Burkitt's lymphoma (Zech et al. 1976), t(14;18) in follicular lymphoma (Yunis et al. 1982), deletion 13q14 in retinoblastoma (Yunis and Ramsey 1978), deletion 11p14 in Wilm's triad syndrome (Miller 1977), and deletion 6q in melanoma (Trent et al. 1983; Becher et al. 1983). Here we wish to describe a specific chromosomal marker, deletion (3)(p14-23), that is associated with and diagnostic of SCLC.

Materials and Methods

Chromosome preparations of bone marrow and short-term (1–2 days culture) or long-term cell lines were made according to previously described methods (Tjio and Whang 1962; Moorhead et al. 1960). Air-dried slides were stained with conventional Giemsa for G-banding analysis using trypsin (Seabright 1971), or for C-banding analysis (Arrighi and Hsu 1971). Detailed chromosome analyses were carried out according to the criteria of the Paris Conference (1971). At least 30 metaphases were examined for breaks or aberrations and 10 G-banded cells karyotyped.

The specimens studied were obtained from patients with histologically confirmed SCLC who were being evaluated by the NCI-VA Oncology Branch (now the Naval Medical Oncology Branch) of the National Institutes of Health, with the exception of cell line NCI-N230, which was derived from the tumor of a Japanese patient (courtesy of Dr. Shimosato of Toyko, Japan) (Shimosato et al. 1979). A total of 3 short-term cultures, 54 direct bone marrow samples, 16 continuous SCLC cell lines, 2 lymphoblastoid lines, and 5 non-SCLC cell lines were examined (Whang-Peng et al. 1982, 1983).

The direct bone marrow specimens are designated by the initial of the patients, continuous cell lines from patient samples have an "H" (human) prefix (NCI-HXXX); those from

Recent Results in Cancer Research. Vol. 97
© Springer-Verlag Berlin · Heidelberg 1985

Table 1. Structural abnormalities in ≥ 50% of metaphases

Patient/ cell line	Sex	Speci- men	Cytol- ogy	Modal #	%A	Abnormalities
FB	M	BM	+?%	45	100	1p+,1p++,+de1(3)(p14−23),14q+
RB	M	BM	+ 60%	70	28	t(1;3)(1q;3q),de1(3)(p14−23),de1(3)(p21q21),de1(22)q
GE	M	BM	+?%	45	90	1p+q21−,t(1;3;9),de1(3)(p11−23),de1(3)(p13q26)
WF	M	BM	+?%	77	43	de1(3)(p14−23),de1(3)(p14−q21),de1(3)(p13)
DF	M	BM	+1%	63	70	de1(3)(p21),de1(3)(p14−23),de1(3)(p14−23q21)
CM	M	BM	−	46	20	de1(3)(p14−23)
JS	M	BM	+50%	77	80	de1(3)(p14q21),t(13;14),de1(22)(q11)
P340		Tumor			80	de1(3)(p14q21),de1(7)(p21),t(13;14),?de1(20)(p11q11),de1(22)(q11)
RT	M	BM	+70%	156	73	de1(3)(p14−23)
HP	F	BM	−	46	0	46,XX
P329		PE				de1(3)(p14−23)
H187	M	PE	SCLC	67		inv(1)(qter→q32::pter→q32),dup(1)(q32−44),de1(3)(p14q13),de1(3)(p14−23),t(7;13)(qter→p11::q11→qter),de1(9)(q11),de1(11)(p11),t(12;19)(pter→qter::pter→qter)
H220	M	PE	SCLC	75		t(2p;3q),de1(3)(p11),de1(3)(p13q13)
P220						t(2p;3q),de1(3)(p11),de1(3)(p13q13)
H69	M	PE	SCLC	40		t(1;16)(qter→q21::qter→pter),t(1;19)(qter→q23::qter→pter),de1(3)(p21−24),de1(3)(p23q26),de1(10)(q22),de1(11)(q23),t(5;13)(qter→q13::q32→pter),de1(17)(p12),DMs
H60	F	PE	SCLC	80		de1(1)(q32),de1(3)(p14q21),t(3;4)(p23→q11::q11→qter),t(5;20)(qter→q11::qter→pter)
H128	M	PE	SCLC	60		de1(3)(p14q23),de1(3)(p14),12q+
H128BL						46,XY
H146	M	BM	SCLC	69		de1(3)(p14q24),t(11;14)(qter→p11::p11→qter),t(11;13)(qter→p11::11→qter),de1(X)(q23)
H209	M	BM	SCLC	49		t(1;1)(qter→q32::p36→qter),de1(3)(p14)
H182	F	LN	SCLC	44		de1(3)(p14−23),rob t(14;14)
H140	M	LN	SCLC	106		de1(1)(p32),de1(3)(p13),de1(3)(p14−23)
H64	M	LN	SCLC	44		de1(1)(q31),t(2p;3q),de1(3)(p14−23),t(9;13)(qter→p11::q11→qter),de1(11)(p15),12q+,de1(X)(q22),min
N230	M	Lung	SCLC	82		de1(3)(p14q13)
H250	M	Brain	SCLC	68		de1(3)(p14−23),de1(22)(q11)
H211	F	BM	SCLC	44		de1(3)(p14−23),t(3;19)(p13→qter::p11→qter),marker #3
H82[a]	M	PE	Large	58		de1(1)(p13),de1(1)(q21),de1(3)(p21),inv(3)(p14−23), HSR on 15p
H196[a]	M	PE	Mixed	110		de1(1)(q41),de1(3)(p13−23),de1(3)(p13),t(3;16)(qter→q24::pter→qter),de1(6)(q24)
H175[a]	M	PE	Mixed		66	de1(3)(p23q26)

[a] Small cell variants

nude mouse heterotransplants have an "N" (nude mouse) prefix (NCI-NXXX); the short-term (2-day) cultures have a "P" (patient) prefix (NCI-PXXX); and those established from B lymphocytes have a BL suffix (NCI-HXXXBL). The 16 SCLC continuous cell lines were derived from one primary tumor, one metastatic brain lesion, 3 metastatic lymph nodes, 3 bone marrows, and 8 pleural effusions (Table 1). The five non-small-cell lines were established from two patients with adenocarcinoma, two with mesothelioma, and one with large cell lung cancer.

The continuous cell lines were established in a serum-free defined medium supplemented with hydrocortisone, insulin, transferrin, 17-β-estradiol, and selenium (HITES) (Carney et al. 1981). Approximately 1×10^7 cells were seeded into 75-cm^2 flasks and incubated at 37° C in 5% CO_2. The cultures were then maintained in RPMI 1640 medium (GIBCO) supplemented with 10% fetal bovine serum. The short-term cultures of the tumor specimens were grown in serum-free growth-factor-supplemented medium.

Continuous cell lines established from SCLC patients are characterized by continuous growth in vitro, the presence of human isoenzymes, formation of colonies in soft agarose, and tumorigenicity in nude mice (Gazdar et al. 1980). Of the 16 lines, 13 have cytological characteristics of SCLC, while the remaining 3 are classed as variants, since their histological characteristics are not typical of SCLC. SCLC cell lines also express a number of biochemical and ultrastructural properties, such as the increased activities of amine precursor uptake and decarboxylation (APUD) cellular enzymes, including L-dopa decarboxylase and neuron-specific enolase. There are high specific activities of creatine kinase BB isoenzymes (radioimmunoassay by Dr. M. Zweig, NIH) and electron microscopy revealed dense-core granules in all SCLC lines with high L-dopa decarboxylase specific activities (Dr. J. Guccion, Washington VA Medical Center) (Gazdar et al. 1980). Non-small-cell lung cancer cell lines lack these biochemical features.

Results

Bone Marrow Samples. Successful cytogenetic studies were possible in 28 of 54 patients. No tumor cells were seen in the histological preparations of 20 patients; 19 of these had normal karyotypes, while 1 (CM) had an abnormal karyotype in 20% of the metaphases. Histologically 8 patients had tumor cells present in the bone marrow, with 7 having an abnormal karyotype; the other (BC) had only normal metaphases noted in the bone marrow. The percentage of aneuploid cells in each of these 8 patients was 100, 28, 90, 43, 70, 20, 60, and 73, respectively. One patient had a modal chromosome number that was hypodiploid; one, diploid; and the remainder had model numbers in the triploid or tetraploid range. Structural abnormalities found in at least 50% of the metaphases are shown in Table 1. All eight patients had a deletion of 3p, with seven having the interstitial deletion 3 (p14-23) in more than 50% of the metaphases; the only exception, patient JS, had this deletion in 35% of the metaphases. Structural abnormalities of chromosome #3 in patient RB are shown in Fig. 1B. Double minutes (DMs) or a homogeneously staining region (HSR) were not seen in the metaphases of any of the direct specimens.

SCLC Cell Lines. Many numerical and structural abnormalities that involved nearly every chromosome in addition to the characteristic structural abnormality involving chromosome #3 (Fig. 2) were found in the 16 SCLC cell lines. There was a wide variation in chromosome numbers in the cell lines, although each had clustered around one or two modal numbers. Hypodiploid lines included NCI-H64, 182, and 211; line NCI-H69 had two

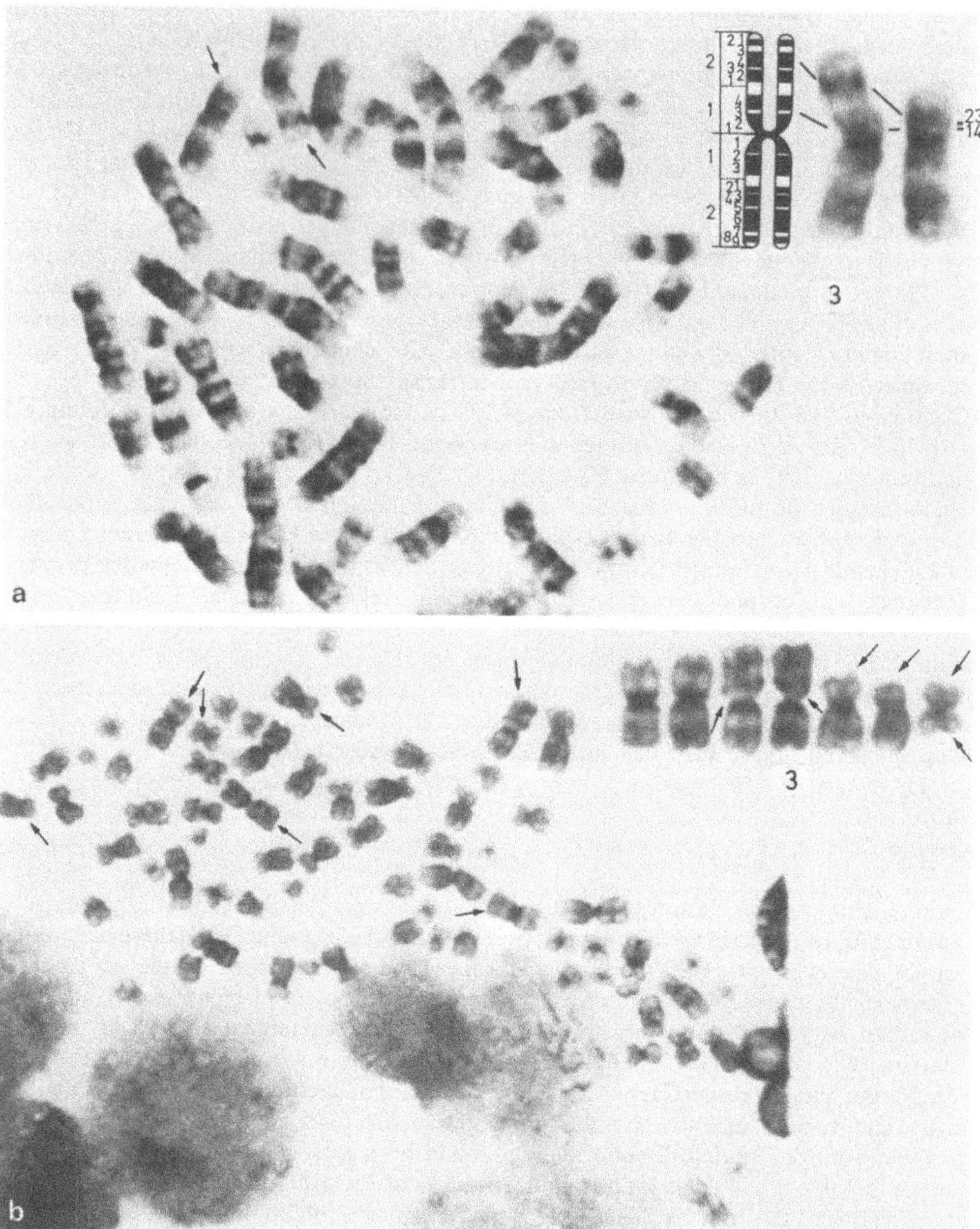

Fig. 1. a G-banded metaphase spread from NCI-H64; *arrows* indicate normal and abnormal #3 chromosomes. *Insert* shows enlarged normal and abnormal #3 chromosomes with the interstitial deletion 3p14-23 along with the idiogram for chromosome #3; **b** G-banded metaphase spread from the direct bone marrow of patient RB; *arrows* indicate normal and abnormal #3 chromosomes. *Insert* shows enlarged #3 chromosomes, both normal and abnormal

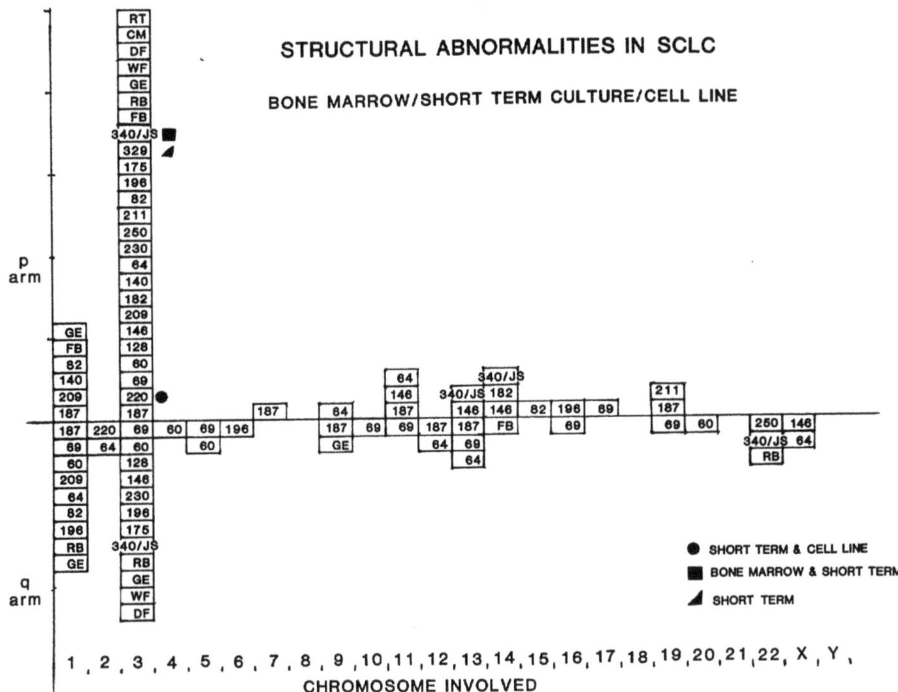

Fig. 2. Structural abnormalities of all the SCLC patients. The abnormalities were found in at least 50% of the abnormal metaphases examined

subpopulations, one being hypodiploid and the second near-tetraploid. The SCLC line NCI-H209 was near-diploid; lines NCI-H82, 175, 187, 128, 146, and 250 had modal chromosome numbers between hyperdiploid and tetraploid; lines NCI-H60, 220, 140, 196, and NCI-N230 were tetraploid; and line NCI-H196 had one mode that was hyperdiploid and a second in the hypertetraploid region. With the exception of the finding that chromosomes #13 and #15 were missing in four of the hypodiploid lines, NCI-H65, 69, 182, and 211, and that an extra #3 chromosome was present in three of these lines, chromosome loss and/or gain appeared to be random.

As expected, the chromosome most frequently involved in structural abnormalities was #3. All of the 16 lines had an abnormality of the short arm of chromosome #3, 3p, in 100% of the metaphases examined. The most frequent abnormality was the interstitial deletion, del(3)(p14-23), which was found in ten of the lines (Fig. 1A). The most bizzare structural abnormalities of #3 were found in NCI-H211, and a morphologically normal #3 was not found; two of these markers consisted of only a small region around the centromere. The chromosome with the next highest frequency of abnormalities was #1, although no specific region of the chromosome seems to be involved. Of the 16 lines, 13 had abnormalities of #1. Structural abnormalities of the rest of the chromosomes found in at least 50% of the metaphases examined are shown in Fig. 1. An HSR was noted in line H-82 and was found on the short arm of chromosome #15 in 96% of the metaphases (Fig. 3A). C-Banding showed this HSR to be euchromatic. DMs were noted in NCI-H60 (20% of the cells) and NIC-H69 (29% of the cells) (Fig. 3B). These DMs ranged in number from 2 per metaphase to very numerous.

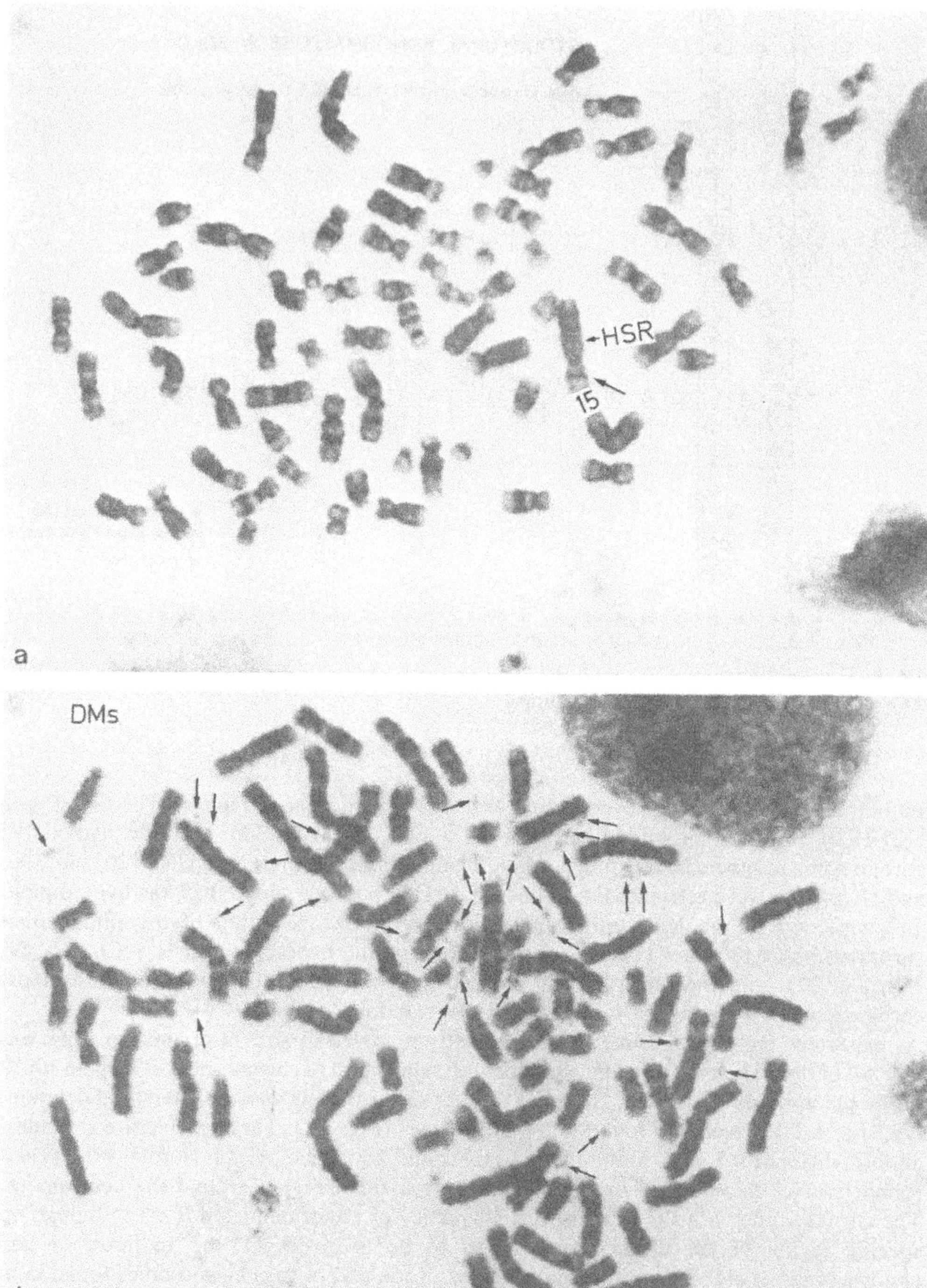

Fig. 3. G-banded metaphases from line H-82 (**a**) with *arrow* indicating the HSR and line H-69 (**b**) with *arrows* indicating the DMs

Chromosomal aberrations were minimal in these lines. NCI-H250 had 10% breaks and NCI-H211 had dicentric marker chromosomes, but in the remainder of the lines no major or minor aberrations were found.

Short-Term Cultures. The chromosomal findings in the short-term cultures were consistent with those found in either the direct bone marrow preparations or the continuous cell lines. The modal number for NCI-P220 was 78, with a range from hyperdiploid to near tetraploid. In addition, the abnormalities of chromosome #3 were identical with those found in the cell line, i.e., del(3)(p14-23), del(3)(p14q13), and t(2p;3q). NCI-P329 had a modal chromosome number of 68, with all the metaphases examined having del(3)(p14-23). NCI-P340 had a modal number of 80 and del(3)(p14q21) in all the metaphases. In addition, 50% of the metaphases had the interstitial deletion del(3)(p14-23), 50% had del(7)(p21), 60% had t(13;14), and 100% had del(22)(q11) and a small metacentric marker that was thought to be del(20)(p11q11). These findings, with the exception of del(7)(p21), are identical with the findings in the direct bone marrow of the patient (JS). Peripheral blood was also examined cytogenetically, and 10% of the metaphases of the 3-day PHA-stimulated culture had an abnormal karyotype matching that of the bone marrow and short-term culture.

Lymphoblastoid Cell Lines. No structural abnormalities of the short arm of chromosome #3 were found in the lymphoblastoid lines examined, NCI-H128BL and NCI-H209BL.

Non-Small-Cell Lung Cancer. None of the cell lines had deletions of 3p below band 3p25, although there was a wide range of abnormalities.

Discussion

Successful cytogenetic studies of specimens from a total of 25 patients with SCLC showed that at least one chromosome #3 in all the metaphases examined had a deletion of the short arm. These included short-term cultures of 1 tumor specimen and 2 pleural effusions, 8 metastatic bone marrow specimens, and 16 long-term SCLC cell lines. Chromosome analysis of lymphoblastoid lines from 2 of these patients and of 5 long-term non-SCLC cell lines showed no deletion of 3p below band p25. The majority of the cell lines (10 of 16) and 7 of 8 bone marrows had the interstitial deletion del(3)(p14-23). The remainder of the specimens also had abnormalities of the short arm of chromosome #3, and shortest region of overlap analysis showed that this portion was affected in all of them. Therefore, the interstitial deletion 3(p14-23) is an acquired defect associated with SCLC and is useful for distinguishing SCLC from other types of lung cancer.

Other interesting findings were observed in the cytogenetic studies of these SCLC patients. Line NCI-H82 had an HSR and lines 64 and 69 had DMs. All these lines were derived from tumors of patients who had relapsed after chemotherapy that included methotrexate. Whether or not the abnormalities in these lines represent gene amplification of such enzymes as dihydrofolate reductase that is associated with resistance to methotrexate remains to be examined. The importance of the HSR and DMs in these cell lines are yet to be determined. Chromosomal abnormalities appear to be very stable, as indicated by studies of NCI-P220 (short-term, 2-day culture) and NCI-H220 (the cell line after 8 months in culture), and of the bone marrow and short-term culture of the tumor in patient JS.

There have been very few reports of cytogenetic studies in SCLC. Studies of involved bone marrow specimens by Wurster-Hill and Maurer (1978) did not specifically mention the deletion of 3p. However, in two of the karyotypes shown a deletion of 3p was noted. The discrepancy in the findings between their report and our studies remains to be resolved. Cell line NCI-N230, derived from a tumor of a Japanese patient, did have an abnormality of 3p, indicating that at least one case of SCLC from Asia had the same deletion. Abnormalities of the short arm of chromosome #3 have been reported in at least one other tumor, viz. renal cell carcinoma. In one family reported by Cohen et al. (1979) there was a translocation between #3 and #8, t(3;8)(p21;q24), which involved a breakpoint at 3p21. In a second report by Pathak et al. (1979) the majority of the metaphases of the direct preparations of the tumor had t(3;11)(p13;15); here the breakpoint was at 3p13. The breakpoint of chromosome #3 in these two reports raises interesting questions concerning the tumorigenicity of this region. A recent report by Luthardt (1982) favors the possibility of a genomic hot spot at 3p14, a region of spontaneous chromosome breaks. However, the low incidence of breaks in the SLCL lines and the direct bone marrow preparations in addition to the high percentage of cells with the deletion of 3p makes the presence of a hot spot at 3p14 less likely.

The development of neoplasia probably requires multiple steps including initiation, promotion, and progression. It is possible that the deletion of 3p is related to chromosomal damage or mutation as a result of exposure to a carcinogen such as cigarette smoke or other chemicals and radiation, and that this specific genetic change eventually leads to SCLC. Gene mapping experiments using rodent-human somatic cell hybrids have shown that the genes required for replication of herpes simplex virus type I are on chromosomes #3 (Francke and Francke 1979) and #11 (Carrit and Goldfarb 1976).

Recently, a remarkable concordance between the chromosomal location of human cellular oncogenes and the breakpoints involved in chromosomal translocations specific for various forms of neoplasms have been described and reviewed by Rowley (1983). Known oncogenes associated with chromosomal abnormalities include: c-*mos* with t(8;21); c-*myc* with t(8;14) and t(8;22); c-*myb* with 6q-, +6, and t(6;14); c-*abl* and c-*sis* with t(9;22); c-*ras*^{H} with 11p-, c-*ras*^{K} with +12; c-*fes* with t(15;17); and c-*src* with 20q-. Studies by Jhanwar et al. (1983) of a set of c-oncogenes belonging to the c-*ras* family, using in situ molecular hybridization techniques, detected three sites with significant hybridization to v-Ki-*ras* and v-Ha-*ras* probes on human pachytene chromosomes. The location of these sites are 11p14.1, 12p21.1, and 12q12.2. In addition, a fourth site located at 3p21.2 exhibited consistent but weak hybridization to both probes. The authors have mapped a c-*ras*-related gene sequence involving chromosomal gain, loss, deletion, duplication, and translocation in a number of myeloid and lymphoid neoplasms with breakpoints at 11p14, 12p11, 12q24, and 3p21. Cytogenetic analysis in tumors of both the families with hereditary renal cell carcinoma and the SCLC patients in our study show a chromosomal abnormality of #3 with a breakpoint in the same vicinity as this oncogene. Translocations or deletions of chromosomes in the regions containing cellular oncogenes may play an important role in the development of or predisposition to neoplasia.

Recent reports by Doolittle et al. (1983), describing the influence of platelet-derived growth factor (PDGF) in normal healing processes as compared with growth of malignant cells, may provide an explanation for changes leading to neoplasia. Amino acid sequences for PDGF were found to be nearly identical with those for c-*sis*, the Simian sarcoma virus. The hypothesis they presented states that the Simian sarcoma virus attacks some cells and inserts the gene for making PDGF. Under normal circumstances PDGF is produced temporarily and causes cells to begin dividing and growing; however, this process ceases when healing has occurred. When the c-*sis* has been incorporated, PDGF production continues and results in growth that may produce a tumor. A similar mechanism may explain the relationship of the c-*ras* gene sequences to the abnormality of chromosome 3p in SCLC, and may be a key element in the cancer-causing events.

References

Arrighi FW, Hsu TC (1971) Localization of heterochromatin in human chromosomes. Cytogenetics 10: 81–86

Becher R, Gibas Z, Sandberg AA (1983) Chromosome 6 in malignant melanoma. Cancer Genet Cytogenet 9: 173–175

Biedler JL, Spengler BA (1976) A novel chromosome abnormality in human neuroblastoma and anti-folate resistant chinese hamster cell lines in culture. J Natl Cancer Inst 57: 683–695

Carney DN, Bunn PA, Gazdar AF, Pagen JA, Minna JD (1981) Selective growth in serum free, hormone supplemented medium of tumor cells obtained by biopsy from patients with small cell carcinoma of the lung. Proc Natl Acad Sci USA 78: 3185–3189

Carrit B, Goldfarb P (1976) A human chromosome determinant for susceptibility to herpes simplex virus. Nature 264: 556–558

Caspersson T, Farber S, Foley GE, Kudynowski J, Mosest EJ, Simonsson E, Wagh U, Zech L (1968) Chemical differentiation along metaphase chromosomes. Exp Cell Res 49: 219–222

Cohen AJ, Li FP, Berg S, Marchetto DJ, Tsai S, Jacobs SC, Brown RS (1979) Hereditary renal-cell carcinoma associated with a chromosomal translocation. N Engl J Med 301: 592–595

Doolittle RF, Hunkapiller MW, Hood LE, Devare SG, Robbins KC, Aaronson SA, Antoniades (1983) Simian sarcoma virus *onc* gene, v-*sis*, is derived from the gene (or genes) encoding a platelet-derived growth factor. Science 221: 275–277

Francke U, Francke BR (1979) Assignment of gene(s) required for herpes simplex virus Type I (HVIS) replication to the long arm of chromosome 11. Human gene mapping 5: fifth international workshop on human gene mapping. Birth Defects: Original Article Series XV: 11, 1979. The National Foundation, New York, p 550

Gazdar AF, Carney DN, Russell EK, Sims HL, Baylin SB, Bunn PA, Guccion JG, Minna JD (1980) Establishment of continuous clonable cultures of small-cell carcinoma of the lung which have amine precursor uptake and decarboxylation cell properties. Cancer Res 40: 3502–3507

Jhanwar SC, Neel BG, Hayward WS, Chaganti RSK (1983) Localization of c-*ras* oncogene family on human germline chromosomes. Proc Natl Acad Sci USA 80: 474–491

Luthardt FW (1982) In vitro induced expression of genomic "hot spot" at 3p14. (abstract) Abstract of the thirty-third annual meeting of the American Society of Human Genetics, p 134A

Miller RW (1977) Cancer and congenital malformation: another view. In: Mulvihill JJ, Miller RW, Fraumeni JF Jr (eds) Genetics of human cancer. Raven, New York, p 81

Moorhead PS, Nowell PC, Mellman WJ, Battips DM, Hungerford DA (1960) Chromosome preparations of leukocytes cultures from human peripheral blood. Exp Cell Res 20: 613–616

Nowell PC, Hungerford DA (1960) A minute chromosome in human chronic granulocytic leukemia. Science 132: 1487

Paris Conference (1971) Standardization in human cytogenetics. Birth Defects: Original Article Series VIII 7 1972. The National Foundation, New York

Pathak S, Strong LC, Ferrell RE, Trindade A (1982) Familial Renal Cell Carcinoma with a 3;11 chromosome translocation limited to tumor cells. Science 217: 939–941

Rowley JD (1983) Human oncogene locations and chromosome aberrations. Nature 301: 290–291

Rowley JD, Golomb HM, Vardiman J, Fukuhara S, Dougherty C, Potter D (1977) Further evidence for a non-random chromosomal abnormality in acute progranulocytic leukemia. Int J Cancer 20: 869–872

Seabright M (1971) A rapid bainding technique for human chromosomes. Lancet II: 971–972

Shimosato Y, Kameya T, Hirohashi F (1979) Growth, morphology, and function of xenotrans planted human tumors. In: Sommers SC, Rosen PT (eds) Pathology annual, part 2. Appleton, New York, pp 215–257

Tjio JH, Whang J (1962) Chromosome preparations of bone marrow cells without prior in vitro culture or in vivo colchicine administration. Stain Technol 37: 17–20

Trent JM, Rosenfeld SB, Meyskens FL (1983) Chromosome 6q involvement in human malignant melanoma. Cancer Genet Cytogenet 9: 177–180

Whang-Peng J, Bunn PA, Kao-Shan CS, Lee EC, Carney DN, Gazdar A, Minna JD (1982) A nonrandom chromosomal abnormality, del 3p(14-23), in human small lung cancer. Cancer Genet Cytogenet 6: 119–134

Whang-Peng J, Carney DN, Lee EC, Kao-Shan CS, Bunn PA, Gazdar A, Minna JD (1983) In: Crispin R (ed) Cancer: etiology and prevention. Elsevier, New York, pp 47–60

Wurster-Hill DH, Mauruer LH (1978) Cytogenetic diagnosis of cancer: Abnormalities of chromosomes and polyploid levels in the bone marrow of patients with small cell anaplastic carcinoma of lung. J Natl Cancer Inst 61: 1065–1075

Yunis JJ, Ramsey N (1978) Retinoblastoma and subband deletion of chromosome 13. Am J Dis Child 132: 161–163

Yunis JJ, Oken MM, Kaplan ME, Ensrud KM, Howe RR, Theologides A (1982) Distinctive chromosomal abnormalities in histologic subtypes of Non-hodgkin's lymphoma. N Engl J Med 307: 1231–1236

Zech L, Haglund U, Nilsson K, Klein G (1976) Characteristic chromosomal abnormalities in biopsies and lymphoid cell lines from patients with Burkitt and non-Burkitt lymphomas. Int J Cancer 17: 47–56

Growth Characteristics and Heterogeneity of Small Cell Carcinoma of the Lung

L. Vindeløv, H. H. Hansen and M. Spang-Thomsen

Medical Department, The Finsen Institute, 49, Strandboulevarden, 2100 Copenhagen, Denmark

Introduction

Cell kinetic data and theories have been valuable in cancer treatment, in that they have provided a theoretical framework for the assessment of treatment results and the formulation of new treatment strategies (Skipper 1971, 1979; Hill 1978; Tannock 1978; Hiddeman 1982).

The available cell kinetic data on small cell carcinoma of the lung (SCCL) are scarce, and have been reviewed extensively (Shackney et al. 1979; Lenhardt et al. 1981; Brigham et al. 1982; Straus et al. 1983). Most of the data were obtained, analysed, and reviewed with the underlying assumption that the tumors are homogeneous. Recent research has indicated that individual malignant tumors, including SCCL, often consist of different malignant subpopulations (Nowell 1976; Vindeløv et al. 1980; Hart and Fidler 1981; Dexter and Calabresi 1982). This heterogeneity, which seems to be a fundamental characteristic of malignant tumors, can profoundly affect their kinetic behavior. In this review we will summarize what is known about the growth kinetics of SCCL, and discuss these data in the context of the heterogeneity of this neoplasm, with the purpose of looking for possible ways of improving the treatment.

Heterogeneity of SCCL

According to the theory of clonal evolution, malignant tumors are monoclonal in origin, i.e., they arise from a single cell. This original tumor cell has become malignant by some unknown process, and has acquired two basic characteristics: It has a growth advantage over the normal cells, and therefore overgrows these, and it is genetically unstable, so that new variants are produced as the population proliferates (Nowel 1976). One implication of this is that tumors may be heterogeneous and contain different, although related, subpopulations of cells. Because of the random nature of the evolution, a further implication is that each individual tumor is unique. There is thus heterogeneity within each tumor and within the population of all SCCL. It is the first type of heterogeneity that is our main concern in this review.

Morphological heterogeneity of SCCL is well established and is taken into account in the WHO subtyping of the tumor. SCCL with foci of large cell tumor constitute about 5%–15% of all SCCL and and are classified as *intermediate*. SCCL with foci of adenocarcinoma or squamous cell carcinoma are classified as *combined* and occur in less than 1% of SCCL at presentation (Minna et al. 1982). The incidence of originally heterogeneous lung tumors is difficult to assess, because of the occurrence of patients with

Recent Results in Cancer Research. Vol. 97
© Springer-Verlag Berlin · Heidelberg 1985

two or more primaries. The published incidence of multiple primaries ranges from 0.5% to 14.5%. The morphology of SCCL at autopsy may have changed after intensive combination chemotherapy, with or without additional radiotherapy. The incidence of exclusively "non-small-cell" carcinoma at autopsy in patients with biopsy-proved diagnosis of SCCL ranges from 5.5% to 12.5%, while the incidence of mixed histologic patterns at autopsy ranges from 14.3% to 32.9% (Abeloff and Eggleston 1981).

A number of other studies indicate that SCCL are heterogeneous as predicted by the theory of clonal evolution. The evidence concerns chromosome constitution, peptide hormone production, and tumorigenicity in nude mice, as reviewed by Minna et al. (1982). Furthermore, evidence of heterogeneity in drug sensitivity was found in two different cell lines established from one patient with SCCL. The cell lines were proven different by their DNA content determined by flow cytometric DNA analysis. They were also different when tested for drug sensitivity after inoculation into nude mice (Engelholm et al. 1982). Clinical in vivo studies based on tumor volume measurements and sequential flow cytometric DNA analysis of fine-needle aspirates have provided evidence of the heterogeneity of individual tumors in DNA content, and of differences in sensitivity to treatment of different subpopulations (Vindeløv et al. 1980, 1982). Heterogeneity in sensitivity to antineoplastic treatment is of particular interest, since treamtnet failure in initially responsive tumors, such as SCCL, is readily explained by the selection and overgrowth of resistant subpopulations during treatment eradicating the sensitive subpopulations. The theory of Goldie and Coldman (1979) is of interest here. It is a mathematical model describing the probability of resistant variant cells being present in a tumor cell population as a function of the mutation rate towards resistance. Not surprisingly, this probability increases with an increasing mutation rate. The interesting prediction of the theory is that for any tumor with a non-zero mutation rate, the likelihood of there being at least one resistant cell will go from a low to a high probability over a very short interval of the tumor's growth. The finding in SCCL that limited disease patients have a better response to treatment and a better prognosis than extensive disease patients is in agreement with the theory of Goldie and Coldman.

Growth Characteristics of SCCL

The growth of a tumor can be described by the Gompertz function (Spang-Thomsen et al. 1980). The increase in cell number is initially exponential, but it is slowed down progressively with time, as the tumor size asymptomatically approaches a maximum value. A mass of 10^9 cells corresponds roughly to 1 g tumor cells, and is considered the smallest tumor burden clinically detectable. This tumor size is reached after 30 doublings of the tumor volume starting from a single cell. The presence of 10^{12} malignant cells corresponds to 1 kg tumor, and is reached after 40 doublings of the volume. Since most patients will die before a tumor volume of 10 kg or 10^{13} cells is reached, the range of tumors seen clinically extends from approximately 10^9 to 10^{12} cells. Since the tumor volume-doubling time (T_d) increases with tumor size, the T_d in the clinical range are longer than those in the subclinical range. The T_d is influenced by the extent of cell loss from the tumor population, usually indicated by the cell loss factor φ, defined as the ratio of the rate of cell loss to the cell production rate (Steel 1967). In addition the T_d is influenced by the cell cycle time T_c, which is the time from one mitosis to the next. The T_c of individual cells, in an otherwise homogeneous population, is known to exhibit substantial variation. Especially in solid tumors, the variation in T_c is so pronounced that the cell kinetics are usually described by

Table 1. Tumor-doubling times in lung cancer (Straus et al. 1983)

Cell type	No. of patients	Mean doubling time (days)	Range (days)
Small cell	63	55	17–264
Large cell	3	92	48–112
Squamous	99	100	7–381
Adenocarcinoma	43	183	17–590

mathematical models in which the most slowly proliferating cells are arbitrarily defined as nonproliferating (Steel 1977). It is important to realize that the T_d can in theory vary within any range if there is a change even in only one of the two factors that affect it, the T_c and φ. In other words, the T_c may be constant and only φ vary or vice versa. Major problems in cell kinetic reasoning arise from the following facts: Direct measurements of T_d can be made only in the clinical range from 10^9 to 10^{12} cells. Below this level only indirect methods are available, such as the "period of risk method" discussed recently by Shackney et al. (1981). The T_c can be determined by the percent labeled mitoses method (PLM) (Steel 1977). The kinetics of tumors are conventionally investigated by this technique, where incorporation of an isotope-labeled DNA precursor is used. However, human malignant tumors are little studied because the necessary biopsy procedures are feasible only in selected patients. Among the human solid tumors studied by PLM, only one was an SCCL (Muggia et al. 1974). Generalizations about this tumor are therefore not justified. Furthermore, the above-mentioned arbitrarily defined nonproliferating cells create problems in interpretion of the results. Finally, estimates of φ are also based on PLM data and should likewise be interpreted with caution. The only hard data available are thus the T_d for the clinical range of growth and estimates of the S-phase size, as mentioned below. T_d for clinically detectable lung tumors are given in Table 1. It is seen that the range is extremely wide, 17–264 days for SCCL, and that the mean is close to 2 months. The overall T_d is thus relatively long for this disease, although some tumors may exhibit rapid growth.

Crowding and poor nutrition are thought to contribute to a slower growth in the clinical range than in the subclinical range (van Putten 1975). This results in a prolongation of T_c and an increase in φ. An indication of how fast SCCL cells are able to grow can be obtained from in vitro experiments and from the study of SCCL heterotransplanted to nude mice. The data of Pettengill and Sorensen (1981) and Sorensen et al. (1981), summarized in Table 2, show a range in vitro of 1.6–35 days. The T_d of the cell lines seem to cluster around 3–6 days or 4–5 weeks. In vivo results in nude mice are available only for the faster-growing cell lines, and a T_d range of 2.9–15.4 days has been recorded. When these results are compared with the clinical data for SCCL (Table 1) it is seen that the T_d from the experimental systems overlap with the clinical results and cover a range of 1.6–264 days.

Table 3 shows the results of calculating the time for 1 cell to multiply to 10^9 cells for different values of φ and T_c, assuming that φ is constant during the growth and using T_c instead of the potential doubling time in the calculations of φ. Available estimates of φ for human solid tumors are in the range of 70%–90% (Tubiana and Malaise 1976). It is seen that similar values for 30 doublings are obtained with a T_c of 55 days and no cell loss and with a T_c of 5 days and $\varphi = 0.90$.

                                                      L. Vindeløv et al.

Table 2. Cell kinetic characteristics of small cell lung cancer cells in vivo and in vitro in nude mice. (Data of Pettengill and Sorensen 1981 and Sorensen et al. 1981)

Cell line	Doubling time in vitro (days)	Doubling time in vivo in nude mice (days)	
		Male recipients	Female recipients
DMS 47	28		
DMS 53	4.9		
DMS 55	4.8		
DMS 79	5.8	6.5	15.4
DMS 92	5.5	10.8	11.7
DMS 114	1.6	7.8	2.9
DMS 139	35		
DMS 148	28−35		
DMS 149	28−35		
DMS 153	5.8	6.1	6.0
DMS 154	35		
DMS 187	4.1	13.1	13.9
DMS 217	28		
DMS 235	4.6	13.1	11.4
DMS 240	7.4		
DMS 273	6.2		

Table 3. Calculations of the time for 1 cell to multiply to give 10^9 cells assuming exponential growth and the simplified formula $T_d = \dfrac{T_c}{1-\varphi}$

T_c (days)	φ	T_d (days)	Time $1-10^9$ cells (30 doublings)
1.6	0	1.6	48 days
5	0	5	5 months
17	0	17	1 year 5 months
30	0	30	2 years 6 months
55	0	55	4 years 7 months
264	0	264	22 years
5	0.5	10	10 months
5	0.90	50	4 years 2 months
5	0.99	500	41 years 8 months

In Fig. 1 we have calculated the number of cell divisions that has to take place to produce a certain number of cells with a given φ. It is seen that with $\varphi = 0.90$, nearly 10^{12} divisions are needed to produce 10^9 cells. This is of interest in the context of the Goldie and Coldman theory (1979). This theory relates the number of resistant phenotypes present to the population size. Since mutations occur as a function of the number of replications of the genome the number of divisions that has taken place seems a more appropriate parameter to indicate the risk of heterogeneity. With a high rate of cell loss even small tumors have a substantial risk of being heterogeneous. The mean T_d of SCCL is smaller than that of the other types of lung cancer (Table 2). The differences are hardly sufficient to allow a kinetic

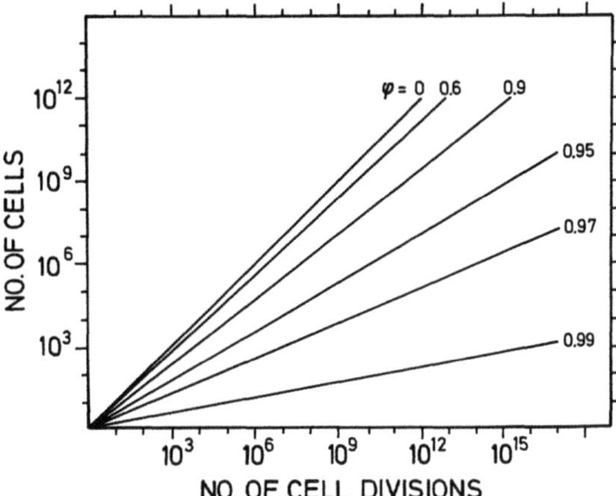

Fig. 1. The relationship between the number of cells present and the number of cell divisions that has taken place in the population was calculated for different values of φ. φ was assumed to remain constant throughout growth

Table 4. Comparison of S-phase[a] fractions of the various types of lung cancer

Cell type	Method	Mean or Median S	Range	No. of Patients	References
Small cell	LI	0.15	0.08 – 0.24	12	Muggia et al. 1974
carcinoma	LI	0.24	0.19 – 0.30	5	Livingston et al. 1974
	LI	0.11	0.02 – 0.28	14	Hainau et al. 1977
	LI	0.12	–	10	Straus et al. 1983
	FCM	0.21	0.14 – 0.43	11	Vindeløv et al. 1982
Large cell	LI	0.10	0.04 – 0.18	5	Muggia 1974
carcinoma	LI	0.11	0.02 – 0.26	29	Hainau et al. 1977
	LI	0.11	–	10	Straus et al. 1983
Squamous cell	LI	0.03	0.03 – 0.04	3	Livingston et al. 1974
carcinoma	LI	0.04	0.01 – 0.10	6	Muggia 1974
	LI	0.08	0.01 – 0.26	38	Hainau et al. 1977
	LI	0.14	–	5	Straus et al. 1983
Adenocarcinoma	LI	0.03	–	2	Livingston et al. 1974
	LI	0.04	0.03 – 0.05	5	Muggia 1974
	LI	0.05	0.01 – 0.21	18	Hainau et al. 1977
	LI	0.02	–	3	Straus et al. 1983

[a] The S-phase were estimated by either [3]HTdR labeling (LI) or by flow cytometric DNA analysis (FCM)

explanation for the differences in responsiveness to therapy, however, and as pointed out by Shackney et al. (1981), SCCL is not a rapidly growing tumor. However, this fact does not exclude the possibility that it is rapidly proliferating. If the cell loss factors for the various types of lung cancer differ substantially the proliferation could do so as well, and more than one would suspect by studying the T_d alone. It is possible, then, that the explanation for the higher responsiveness to therapy of SCCL is based on rapid proliferation, which in general is correlated to drug sensitivity (Tannock 1978). Similar thoughts have been put forward by Shackney et al. (1981). The hypothesis is supported by data for S-phase sizes in lung cancer (Table 4). The S-phase fraction is considered a rough indicator of proliferative activity.

Table 5. State-of the-art results of treating SCCL

Extent of disease	CR	PR	Median survival (months)	Lomg-term survival (> 3 years)
Limited disease	50%	40%	14	15% − 20%
Extensive disease	20%	40%	7	3% − 5%

Heterogeneity, Cell Kinetics, and Possible Implications for Treatment

The emerging picture of a fully developed clinically detectable SCCL as consisting of several different subpopulations of cells has implications for treatment. Some are well known and some have so far received little attention, but they could possibly lead to therapeutic progress if taken into account.

At present, the results of intensive combination chemotherapy of patients with SCCL are as outlined in Table 5 (Aisner et al. 1983). Although it is initially a highly responsive tumor, most patients suffer a relapse which cannot be controlled by subsequent therapy due to resistance of the relapsing tumors.

Combination chemotherapy is superior to treatment with single agents. This has been explained by a potentiating effect of the different drugs involved, and synergism may still be important. However, with the knowledge that the tumors are heterogeneous, an obvious advantage of using more than one drug is that the cells resistant to drug A could be killed by drug B. The emphasis in scheduled design is thus shifted from timing of the individual drugs to the use of non-cross-resistant drugs with activity against SCCL cells (Vindeløv et al. 1982). New drugs with these characteristics have a strong potential for improving the therapy, and a search for such compounds should therefore have a high priority. The fact that treatment with non-cross-resistant drug combinations so far has not significantly improved the results (Aisner et al. 1983) does not prove that the principle is wrong. It is quite possible that the subpopulations in SCCL that survive first-line treatment are generally unresponsive to all the cytostatic agents known at present. Hence the need for new drugs.

The finding that SCCL may contain subpopulations of other histological types of lung cancer and that patients with SCCL initially may show a complete or partial change in histology at autopsy is of interest here. Since the other types of lung cancer are generally unresponsive to treatment, it seems fair to presume that such changes do in fact represent the selection and overgrowth of a subpopulation generally resistant to therapy. What remains to be clarified is the reason for this unresponsiveness. As far as unresponsive experimental tumors are concerned, there is little doubt that in many cases they are permanently and biochemically resistant to the drugs in question. Another possibility is that the resistance is kinetically based. Slowly proliferating tumors are generally less sensitive to antinoeplastic treatment than fast proliferating tumors, and it is well known that all human tumors which are curable today by chemotherapy are quite rapidly proliferating (Shackney et al. 1978). Again, a tumor that was initially "pure" SCCL, but is found at autopsy to have changed to some other type, could be taken as an example of a kinetic change, since the other types of lung cancer proliferate more slowly than SCCL (Table 4). A kinetic change could also take place without a morphological change, however. The data of Pettergill and Sorensen (1981) and Sorensen et al. (1981) on in vitro T_d for cell lines (Table 3) show that the values cluster around 3−6 days and 4−5 weeks.

Considering the effect of changes in φ and T_c such a change from fast to slow proliferation could take place without any change in the actual T_d of the tumor, and thus without any clinically detectable change. Further and more direct evidence of a kinetic change comes from sequential flow cytometric DNA analysis of SCCL during treatment. A decreased S-phase fraction of some tumors relapsing during chemotherapy was demonstrated (Vindeløv et al. (1980). In these cases the treatment might have caused a preferential kill of the rapidly proliferating cells, and thus facilitated the subsequent overgrowth of the slowly proliferating subpopulation.

As a consequence of the data and the speculations outlined above we feel the possibility that the resistance of SCCL is kinetically based should be considered, and also rational ways of treating such slowly proliferating subpopulations. It seems paradoxical that slowly proliferating tumors are the most difficult to treat. Although their sensitivity is smaller the cell production is also smaller, as is the cell kill needed to induce tumor regression.

We would expect slowly proliferating cells to react to a chronic treatment, providing low concentrations of drug for days or weeks, rather than high concentrations for minutes or hours. We would also expect the responses to be less dramatic, i.e., a slower tumor regression than is usually seen, because the tumor regression would also be limited by the longer cell-cycle transit times.

In conclusion, we think that treatment approaches designed at hitting slowly proliferating subpopulations of SCCL, and given after an initial induction treatment of the kind used today, might have the potential for changing the treatment results, which have reached a plateau in recent years.

Acknowledgement. The authors wish to thank Ib Jarle Christensen for performing the calculations on which Fig. 1 is based.

References

Abeloff MD, Eggleston JC (1981) Morphologic changes following therapy. In: Greco FA, Oldham RK, Bunn PA Jr (eds) Samm cell lung cancer. Grune and Stratton, New York, pp 235–259

Aisner J, Alberto P, Bitran J, Comis R, Daniels J, Hansen HH, Ikegami H, Smyth J (1983) Role of chemotherapy in small cell lung cancer: A consensus report of the International Association for the Study of Lung Cancer Workshop. Cancer Treat Rep 67: 37–43

Brigham BA (1982) Small cell anaplystic carcinoma of the lung. A review of growth characteristics and implications for chemotherapy. Cancer Chemother Pharmacol 9: 1–5

Dexter DL, Calabresi P (1982) Intraneoplastic diversity. Biochim Biophys Acta 695: 97–112

Engelholm SA, Spang-Thomsen M, Vindeløv LL, Nielsen A, Hansen HH (1982) Different sensitivity to antineoplastic therapy of two subpopulations of a single human small cell carcinoma of the lung. Abstract of the IIIrd world conference on lung cancer, Tokyo

Goldie JH, Coldman AJ (1979) A mathematic model for relating the drug sensitivity of tumors to their spontaneous mutation rate. Cancer Treat Rep 63: 1727–1733

Hainau B, Dombernowsky P, Hansen HH, Borgeskov S (1977) Cell proliferation and histologic classification of bronchogenic carcinoma. J Natl Cancer Inst 59: 113–118

Hart IR, Fidler IJ (1981) The implications of tumor heterogeneity for studies on the biology and therapy of cancer metastasis. Biochim Biophys Acta 651: 37–50

Hiddeman W, Büchner T, Andreef M, Wörmann B, Melamed M, Clarckson BD (1982) Cell kinetics in acute leukemia. A critical reevaluation based on new data. Cancer 50: 250–258

Hill BT (1978) The management of human solid tumours: Some observations on the irrelevance of traditional cell cycle kinetics and the value of certain recent concepts. Cell Biol Int Rep 2: 215–230

L. Vindeløv et al.

Lenhard RE Jr, Woo KB, Freund JS, Abeloff MD (1981) Growth kinetics of small cell carcinoma of the lung. Eur J Cancer Clin Oncol 17:899–904

Livingston RB, Ambus U, George SL, Freireich EJ, Hart JS (1974) In vitro determination of thymidine-^{3}H labeling index in human solid tumours. Cancer Res 34:1376–1380

Minna JD, Carney DN, Alvarez R, Bunn PA Jr, Cutttitta F, Ihde DC, Matthews MJ, Oie H, Rosen S, Whang-Peng J, Gazdar AF (1982) Heterogeneity and homogeneity of human small cell lung cancer. In: Owens AH, Coffey DS, Baylin SB (eds) Tumor cell heterogeneity. Origins and implications. Academic, New York, pp 29–52

Muggia FM (1974) Cell kinetic studies in patients with lung cancer. Oncology 30:353–361

Muggia FM, Krezoski SK, Hansen HH (1974) Cell kinetic studies in patients with small cell carcinoma of the lung. Cancer 34:1683–1690

Nowel PC (1976) The clonal evolution of tumor cell populations. Science 194:23–28

Pettengill OS, Sorensen GD (1981) Tissue culture and in vitro characteristics. In: Greco FA, Oldham RK, Bunn PA Jr (eds) Small cell lung cancer. Grune and Stratton, New York, pp 51–77

Shackney SE, McCormack GW, Cuchural GJ Jr (1978) Growth rate patterns of solid tumors and their relation to responsiveness to therapy. Ann Intern Med 89:107–121

Shackney SE, Cohen MH, Bunn PA Jr, Ihde DC, Minna JD (1979) The application of principle of cell kinetics in the design of treatment regimens for small cell carcinoma of the lung. In: Muggia FM, Rozencweig M (eds) Lung cancer: Progress in therapeutic research. Raven, New York, pp 63–71

Shackney SE, Strauss MJ, Bunn PA Jr (1981) The growth characteristics of small cell carcinoma of the lung. In: Greco FA, Oldham RK, Bunn PA Jr (eds) Small cell lung cancer. Grune and Stratton, New York, pp 225–234

Skipper HE (1971) The cell cycle and chemotherapy of cancer. In: Baserga R (ed) The cell cycle and cancer. Dekker, New York

Skipper HE (1979) Historic milestones in cancer biology: A few that are important in cancer treatment (Revisisted). Semin Oncol 6:506–514

Sorensen GD, Pettergill OS, Cate CC (1981) Studies on xenografts of small cell carcinoma of the lung. In: Greco FA, Oldham RK, Bunn PA Jr (eds) Small cell lung cancer. Grune and Stratton, New York, pp 95–121

Spang-Thomsen M, Nielsen A, Visfeldt J (1980) Growth curves of three human malignant tumors transplanted to nude mice. Exp Cell Biol 48:138–154

Steel GG (1967) Cell loss as a factor in the growth rate of human tumors. Eur J Cancer 3:381–387

Steel GG (1977) Growth kinetics of tumours. Clarendon, Oxford

Straus MJ, Moran RE, Shackney SE (1983) Growth characteristics of lung cancer. In: Straus MJ (ed) Lung cancer. Clinical diagnosis and treatment. Grune and Stratton, New York, pp 63–84

Tannock I (1978) Cell kinetics and chemotherapy: A critical review. Cancer Treat Rep 62:1117–1133

Tubiana M, Malaise EP (1976) Growth rate and cell kinetics in human tumors: some prognostic and therapeutic implications. In: Symington T, Carter KL (eds) Scientific foundations of oncology. William Heinemann Medical Books, London, pp 126–136

van Putten LM (1975) Problems in the treatment of slow-growing tumors. In: Grundmann E, Gross R (eds). The ambivalence of cytostatic therapy. Springer, Berlin Heidelberg New York, pp 225–233

Vindeløv LL, Hansen HH, Christensen IJ, Spang-Thomsen M, Hirsch FR, Hansen M, Nissen NI (1980) Clonal heterogeneity of small-cell anaplastic carcinoma of the lung demonstrated by flow-cytometric DNA analysis. Cancer Res 40:4295–4300

Vindeløv LL, Hansen HH, Gersel A, Hirsch FR, Nissen NI (1982) Treatment of small-cell carcinoma of the lung monitored by sequential flow cytometric DNA analysis. Cancer Res 42:2499–2505

Whang-Peng J, Kao-Shan CS, Lee EC, Bunn PA Jr, Carney DN, Gazdar AF, Minna JD (1982) Specific chromosome defect associated with human small-cell lung cancer: Deletion 3p(14-23). Science 215:181–282

In Vitro Studies in Small Cell Lung Cancer Cell Lines

C. Gropp, W. Luster, and K. Havemann

Abteilung Hämatologie/Onkologie, Zentrum für Innere Medizin der Universität Marburg,
Baldingerstrasse, 3550 Marburg, Federal Republic of Germany

Introduction

The establishment of permanent cell lines from small cell lung tumors (SCLC) in several
laboratories (Luster et al. 1983; Pettengill et al. 1980; Gazdar et al. 1980; Sorenson et al.
1981) has greatly increased our knowledge of the biology of SCLC in recent years. The in
vitro studies performed with these cell lines have indicated that SCLC may be of
neuroendocrine origin and is closely related to bronchial endocrine cells of the K cell type.
The cell lines derived from SCLC share a number of characteristics with the K cells. As has
been shown for K cells, cultured SCLC cells express amine precursor uptake and
decarboxylation (APUD) properties, as L-dopa decarboxylase, dense-core granules,
formaldehyde-induced fluorescence, neuron-specific enolase, bombesin, and other peptide
hormones are all found (Baylin et al. 1980; Marangos et al. 1982; Moody et al. 1981).
Beside APUD cell properties, in vitro studies on SCLC cultures revealed other
characteristics by which this tumor may be distinguished from non-small-cell lung cancer
(NSCLC). Cytogenetic studies in SCLC tumors and cell lines have detected a characteristic
chromosomal abnormality deletion 3p(14-23) in 100% of the cells (Whang-Peng et al.
1982). This abnormality does not seem to be associated with NSCLC.
In recent years a number of peptide hormones have been detected in sera of patients with
SCLC, but also in the medium of cell lines of this tumor (Sorenson et al. 1981; Wolfsen et
al. 1979; Luster et al. 1983; Ratcliffe et al. 1982; Ellison et al. 1973; Gropp et al. 1980,
1982). Sometimes elevated peptide hormone levels are associated with paraneoplastic
syndromes, such as Cushing's syndrome or the syndrome of inadequate arginine
vasopression (AVP) secretion. Recent studies show that peptide hormones, mostly ACTH,
calcitonin, and AVP are elevated in the sera of two-thirds of patients with SCLC. In
contrast, only 3%−5% of these patients clinically have a paraneoplastic syndrome.
Therefore the biochemical nature of these hormones, which may be biologically inactive
precursor hormones, is of great interest. As these hormones are secreted in large amounts
into the medium by cultured cell lines we are interested in our laboratory in establishing a
number of hormone-producing lung cancer cell lines as a source of further biochemical
investigations. We are further interested in the possible pathophysiological role of these
hormones for tumor growth or tumor cell differentiation.

Establishment of Lung Tumor Cell Lines

Tumor cell lines were established from surgical biopsies, from biological fluids (pleura,
pericardium), or from bone marrow metastases. In brief, solid tissue specimens were

Recent Results in Cancer Research. Vol. 97
© Springer-Verlag Berlin · Heidelberg 1985

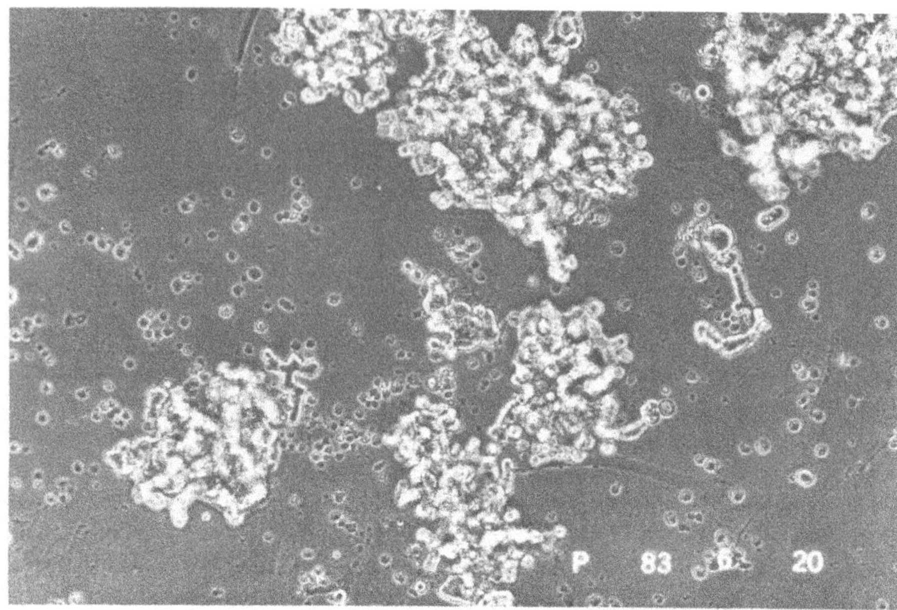

Fig. 1. Phase contrast photomicrograph of a small cell cancer cell line growing as floating aggregates

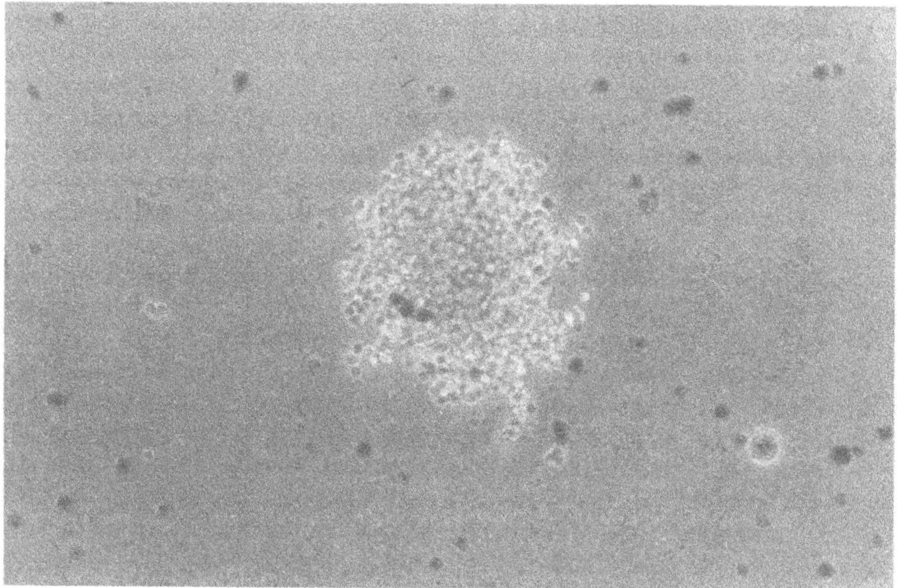

Fig. 2. Colonies of a small cell lung cancer cell line growing in soft agar

Table 1. Characterization of long-term lung cancer cell cultures

Lung cancer cell line		Tissue of origin	Sex	Doublingtime	Hormones produced
Small cell carcinoma	MR-16	Bone marrow aspirate	M	3 days	Bombesin
	MR-22	Pleural fluid	M	2 days	Calcitonin, bombesin
	MR-26	Bone marrow aspirate	M	20 days	ACTH
	MR-27	Pleural fluid	M	15 days	Bombesin
	MR-28	Bone marrow aspirate	M	10 days	ACTH, calcitonin, bombesin, PTH, substance p
	MR-55	Lung biopsy	M	5 days	ACTH, calcitonin, bombesin, neurotensin
Squamous cell carcinoma	MR- 9	Lung biopsy	M	1.5 days	ACTH, bombesin
	MR-20	Lung biopsy	M	3 days	ACTH, substance P
	MR-25	Lung biopsy	M	1 day	ACTH, calcitonin, bombesin, neurotensin
	MR-32	Lung biopsy	M	1.5 days	ACTH, calcitonin, bombesin
	MR-33	Lung biopsy	M	−0.5 days	ACTH, calcitonin, bombesin, PTH
	MR-50	Lung biopsy	M	2 days	ACTH, calcitonin, bombesin
	MR-65	Lung biopsy	M	1 day	ACTH, calcitonin, bombesin, substance P
Adeno-carcinoma	MR- 4	Lung biopsy	M	4 days	ACTH
	MR- 5	Lung biopsy	M	1−3 days	ACTH
	MR-13	Lung biopsy	M	6 days	ACTH, calcitonin, bombesin, neurotensin
Large cell carcinoma	MR- 2	Lung biopsy	M	10 days	ACTH, calcitonin, bombesin, substance P
	MR- 7	Lung biopsy	M	2 days	ACTH, calcitonin, bombesin, PTH
	MR- 8	Lung biopsy	M	3 days	ACTH, calcitonin, bombesin, PTH
	MR-58	Lung biopsy	M	5 days	ACTH, neurotensin, bombesin

washed with antibiotics and minced into 1- to 3-mm^3 pieces. To obtain a cell suspension the material was desintegrated with collagenase and sometimes additionally with trypsin. Fluid specimens were collected with an anticoagulant, centrifuged, and then separated from erythrocytes and cell debris by Ficoll gradient centrifugation. After these procedures tumor cells were diluted to 1×10^5 cell/ml medium (M&M Dulbecco with 16.6% fetal calf serum) and plated into 2-cm^2 petri dishes. After 5–20 days the medium was assayed for hormone content and the positive cell lines were first subcultured in microtiter plates and then established in 25-ml plastic flasks (Gropp et al. 1983; Luster et al. 1983). In addition, cloning was performed by the soft agar cloning technique (Figs. 1 and 2). Tumor cell lines were also established after heterotransplantation of lung tumors in nude mice (NMRI nu/nu). Up to now 97 lung tumors have been cultured for 5–20 days. After this time the medium is assayed for peptide hormones, and 20 hormone-producing lung tumor cell lines of all types of histology have been established as long-term cultures and as heterotransplants in nude mice. The characteristics of these cell lines, with reference to electron microscopy, immunohistology, growth characteristics, and cytogenetic studies will be described in detail elsewhere. Table 1 shows some examples of hormone-producing cell lines established in our laboratory. It has to be stressed that besides cell lines from SCLC, cell cultures from NSCLC also secrete a variety of peptide hormones into the medium. Because we are interested in the further characterization of these peptide hormones, and especially in ACTH and calcitonin, we have also established six permanent C cell carcinoma cell lines that also produce high amounts of calcitonin and in some cases ACTH, bombesin, and neurotensin (Luster et al. 1982).

Isolation and Characterization of Peptide Hormones

Calcitonin-Immunoreactive Protein

Recently we described the separation of three calcitonin-immunoreactive proteins from sera of patients with small cell lung cancer by means of gel filtration techniques (Luster et al. 1982). These proteins, with molecular weights of 100,000, 48,000, and 20,000 daltons, were degradable by incubation with sodium dodecyl sulfate under reducing conditions to a 17,000-dalton protein; this is relatively stable and might be a calcitonin prohormone synthesized by the tumor cells.

To confirm these results and for further characterization of the tumor calcitonin we started in vitro studies with calcitonin-producing lung tumor cell lines.

Calcitonin-containing culture medium was lyophilized and subjected to gel chromatography on AcA 54 (LKB Stockholm, Sweden) columns. By this method three calcitonin-immunoreactive protein fractions with molecular weights of 100,000, 50,000, and 20,000 daltons were identified in the medium of SCLC, but also of NSCLC, cell lines (Fig. 3). These high-molecular-weight calcitonin fractions correspond to the calcitonin proteins isolated in former studies from sera of patients with SCLC. There was no difference between calcitonin proteins from SCLC and NSCLC cultures, and no protein with the molecular weight of physiological calcitonin was detectable.

In further experiments the different calcitonin fractions were incubated in the presence of sodium dodecyl sulfate and then separated by electrophoresis on 10% polyacrylamide gels. The denaturation of the various fractions resulted in a 17,000-dalton calcitonin-immunoreactive band. An additional 3,400-dalton calcitonin appeared on the gels, which has a molecular weight similar to that reported for physiological calcitonin (Fig. 4).

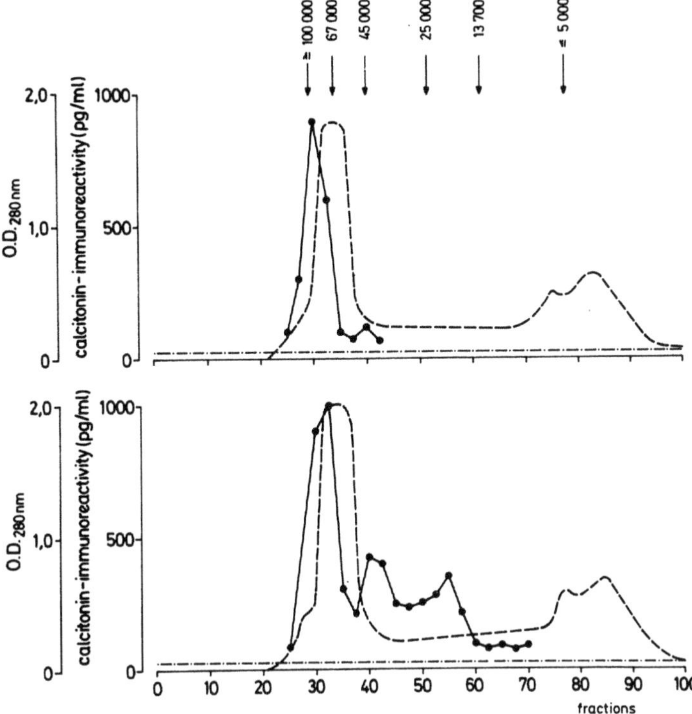

Fig. 3. AcA 54 column chromatography of a small cell lung cancer culture medium (*lower part*) and of an adenocarcinoma culture medium (*upper part*). Fractions were assayed for calcitonin

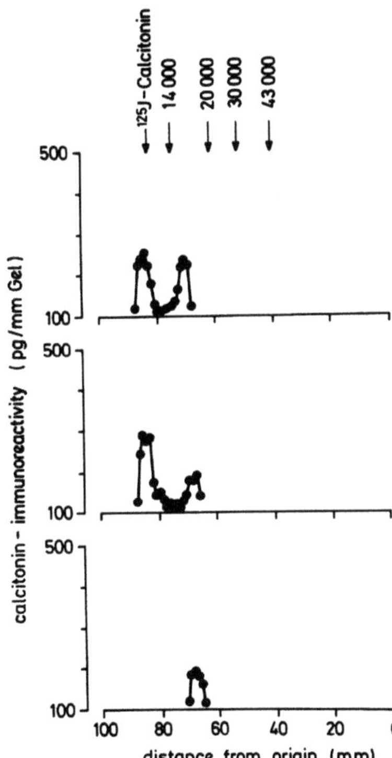

Fig. 4. SDS polyacrylamide gel electrophoresis of the three calcitonin fractions obtained by gel filtration from highest to lowest molecular weight (see Fig. 3). Gels were cut into 1-mm pieces and calcitonin was estimated

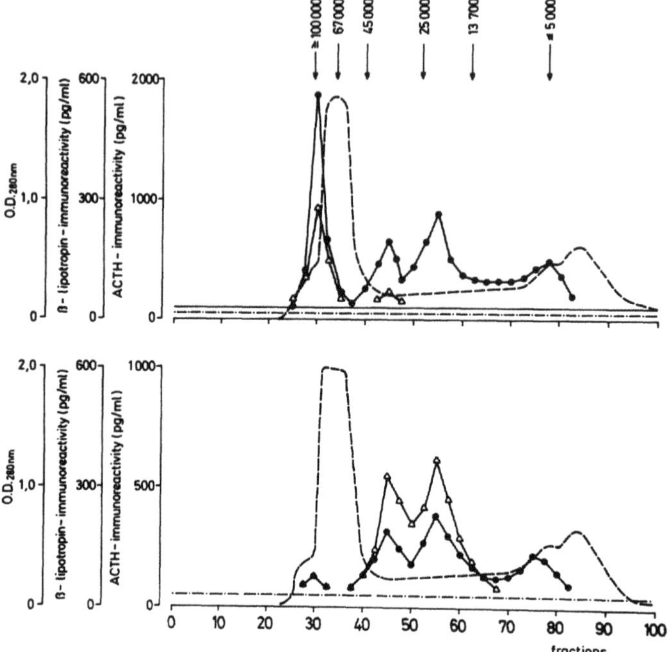

Fig. 5. AcA 54 column chromatography of a small cell lung cancer culture medium (*lower part*) and a large cell culture medium (*upper part*). Fractions were assayed for ACTH and β-lipotrophin

The calcitonin-immunoreactive proteins were further characterized by their behavior on ion-exchange chromatography. Most of the 20,000-dalton calcitonin fraction was bound to DEAE-Sephacel at neutral pH and low conductivity and could be eluted at slightly acid pH in the presence of 55 mM NaCl. The 100,000-dalton fraction did not interact with DEAE-Sephacel.

Affinity chromatography on concanavalin A-Sepharose and lentil lectin-Sepharose showed that the calcitonin-immunoreactive proteins did not contain any glycoprotein component with α-D-mannosyl or sterically related residues. To investigate the stability of the calcitonin fractions against proteolytic degradation they were incubated for different times in the presence of trypsin. This procedure resulted in a degradation of the 100,000- and 50,000-dalton calcitonin-immunoreactive protein to a relatively stable 17,000-dalton protein. This 17,000-dalton protein seems to be a relatively stable core-protein, which may represent the calcitonin prohormone synthesized by the lung tumor cells. Preliminary studies with C cell carcinoma cell lines indicate that the calcitonin-immunoreactive protein from lung cancer cell lines is biochemically different from the calcitonin synthesized by C cell carcinoma cells (Luster et al., to be published).

ACTH-Immunoreactive Protein

In patients' sera the ACTH levels are relatively low and the ACTH is highly sensitive to proteolytic activities. Therefore, direct separation of ACTH-immunoreactive material from sera by gel filtration is impossible. For isolation and characterization of this peptide hormone it is beneficial to establish hormone-producing cell lines. In addition, we have

Table 2. Characterization of calcitonin and ACTH immunoreactive proteins from serum and tissue of lung tumor patients

Immuno-reactivity	Molecular weight determined by gel filtration	Protein A affinity chromatography	Lectinchromatography
Calcitonin	100,000 48,000 20,000	No interaction with immunoglobulins	No glycoprotein component
ACTH	100,000 30,000 20,000 4,800	Binding of IgG	Glycoproteincomponent

Immuno-reactivity	Ion exchange chromatography	SDS stability	Stability against proteolytic activities	Isoelectric point
Calcitonin	Specific enrichment of the 20,000-dalton fraction	Degradation to 17,000-dalton core protein	17,000-dalton protein most stable	5.5−6.0
ACTH	No interaction	No longer measurable by immunological methods	Labile	−

devised a method for further enrichment of the ACTH-immunoreactive material by way of affinity chromatography on Cibacron Blue F3GA followed by lyophilization. After this step, gel filtration resulted in four ACTH-immunoreactive peaks, of more than 100,000, 30,000, 20,000, and 4,500 daltons (Fig. 5). The two major peaks also showed immunoreactivity with β-lipotrophin, which probably derives from the same precursor molecule as ACTH (Gropp et al. 1983). Our results also show that like calcitonin, ACTH is also synthesized both from SCLC and from NSCLC cell lines. After the establishment of several ACTH-secreting lung tumor cell lines and methods for the isolation and enrichment of this peptide hormone (Luster et al. 1982), further studies for the characterization of ACTH-immunoreactive proteins are under way. Table 2 gives a brief summary of the characteristics of the calcitonin and ACTH synthesized by our cell lines.

Influence of Peptide Hormones on Tumor Cell Proliferation

The role of the peptide hormones for growth and function of the hormone-producing cells themselves, i.e., an autocrine or paracrine role, is of great interest. For example, a lung tumor cell line, BEN, has been described, which produces calcitonin and has calcitonin receptors (Ham et al. 1980). In our laboratory we started investigations of the influence of physiological human hormones on the proliferation and biosynthetic activities of lung tumor cells. In these experiments small cell and non-small-cell lung tumor cultures were

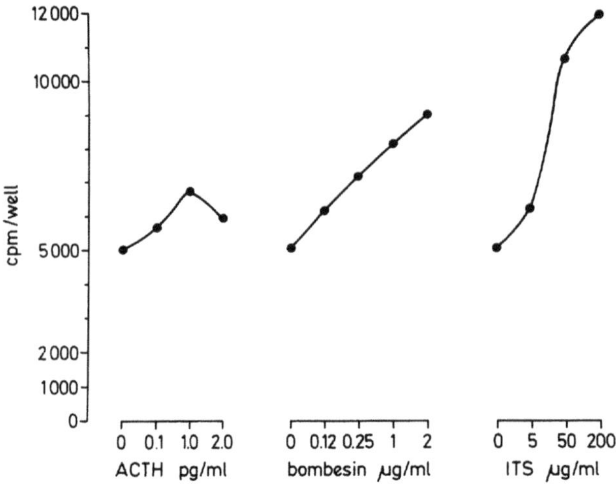

Fig. 6. Influence of peptide hormones and other biological active substances (*I*, insulin; *T*, transferrin; *S*, selenium) on the proliferation of lung tumor cells in vitro

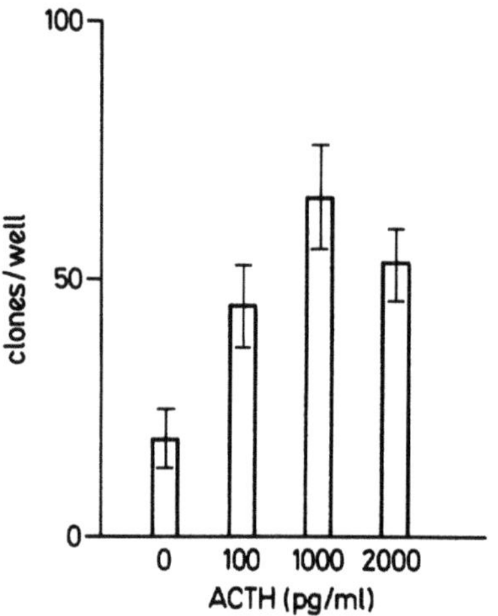

Fig. 7. Influence of ACTH on the cloning efficiency of a small cell lung cancer cell line

incubated in the presence of hormone-containing medium, and cell proliferation was measured by incorporation of radiolabeled thymidine in to the DNA. The results are shown in Fig. 6. Not only the proliferation but also the hormone production of some cell lines was increased by incubation of the cultures with human hormones (Luster et al. 1983). In other experiments the growth of cell lines in soft agar could be influenced by some hormones. The number of colonies of a small cell lung cancer line could be significantly increased by the addition of ACTH (Fig. 7). These results suggest an autocrine or paracrine role of the peptide hormones synthezised by the tumor cells in vitro.

Conclusions

In recent years we have established a number of hormone-producing cell lines from SCLC. These cell lines can secrete various peptide hormones into the culture medium simultaneously. Whether the different hormones are produced by different cells in a heterogeneous cell population is not clear. This question will possibly be answered by cloning studies. Beside SCLC cell lines, several cell lines from NSCLC have been cultured in our laboratory. These cell lines also produce various peptide hormones and no difference in the hormone profile has been detected in comparison with the SCLC cell lines. For example, we have three ACTH-producing adenocarcinomas, and the cell line with the highest calcitonin secretion is also well characterized as an adenocarcinoma. In addition, bombesin, a peptide hormone believed to be highly specific for SCLC, has been demonstrated in some squamous and adenocarcinoma cell lines. These results suggest that lung tumors of all histological types are able to synthesize peptide hormones. This hypothesis is also supported by earlier immunohistological studies on lung tumor tissues, in which we were able to demonstrate ACTH, calcitonin, and β-lipotrophin in nearly 50% of NSCLC tumors (Gropp et al. 1981). Our studies on the isolation and characterization of calcitonin and ACTH secreted from lung cancer tumor cells showed a relatively stable calcitonin-immunoreactive protein with a molecular weight of 17,000 daltons and ACTH-immunoreactive proteins with molecular weights of more than 100,000, 30,000, 20,000, and 4,500 daltons.

These results lead to the conclusion that lung tumor cells synthesize an ACTH and a calcitonin prohormone, which are secreted by the tumor cells after incomplete intracellular degradation. Finally, cell proliferation of SCLC and NSCLC cell lines have been stimulated in vitro by peptide hormones. Further studies will show whether these peptide hormones might act as growth factors for lung tumor cells.

References

Baylin SB, Abeloff MC, Goodwin G et al. (1980) Activities of L-dopa decraboxylase and diamine oxidase (histaminase) in human lung cancers and decarboxylase as a marker for small (oat) cell cancer in cell cultures. Cancer Res 40: 1990−1994

Ellison M, Woodhouse D, Hillyard C et al. (1973) Immunoreactive calcitonin production by human lung carcinoma cells in culture. Br J Cancer 32: 373−379

Gazdar AF, Carney DN, Russel EK et al. (1980) Establishment of continuous clonable cultures of small-cell carcinoma of the lung which have amine precursor uptake and decarboxylation cell properties. Cancer Res 50: 3502−3507

Gropp C, Sostmann H, Luster W, Kalbfleisch H, Lehmann FG, Havemann K (1981) ACTH, β-lipotropin, β-endorphin, β-HCG, Calcitonin and CEA in lung tumor tissues. In: Uhlenbruck C, Wintzer G (eds) CEA und andere Tumormarker. Tumordiagnostik-Verlag, Leonberg, pp 217−226

Gropp C, Havemann K, Kalbfleisch H, Luster W, Sostmann H (1982) Antidiuretisches Hormon bei Patienten mit Bronchialkarzinom. Dtsch Med Wochenschr 107: 977−980

Gropp C, Luster W, Havemann K (1983) High molecular adrenocorticotropin and calcitonin immunoreactive proteins as biosynthetic products of small-cell lung tumors. Acta Endocrinol [Suppl] (Copenh) 253: 16−17

Gropp C, Luster W, Havemann K, Wahl R, Röher HD (1984) Lung and gastrointestinal tumor cells secrete peptide hormones. In: Peeters H (ed) Protides of the biological fluids, vol 31. Brussels, pp 599−602

Gropp C, Havemann K, Scheuer A (1980) Ectopic hormones in lung cancer patients at diagnosis and during therapy. Cancer 46: 347−354

Ham J, Ellison ML, Lumsden J (1980) Tumor calcitonin: interaction with specific calcitonin receptors. Biochem J 190: 545−550

Luster W, Gropp C, Sostmann H, Kalbfleisch H, Havemann K (1982) Demonstration of immunoreactive calcitonin in sera and tissues of lung cancer patients. Eur J Cancer Clin Oncol 18: 1275−1283

Luster W, Gropp C, Havemann K (1983) Peptide hormone synthesizing lung tumor cell lines: establishment and first characterization of biosynthetic products. Acta Endocrinol [Suppl] (Copenh) 253: 24–25

Luster W, Gropp C, Havemann K, Wahl R, Röher HD (to be published) Permanente C-Zell Karzinomlinien und ihre Sekretionsprodukte

Luster W, Gropp C, Loeck MR, Havemann K (1984) Modification of tumor cell proliferation and peptide hormone secretion. In: Peeters H (ed) Protides of the biological fluids, vol 31. Pergamon, Brussels, pp 727–730

Luster W, Gropp C, Havemann K (to be published) Adrenocorticotropin, β-Lipotropin-immunoreactive Proteine – Ein Sekretionsprodukt von Lungentumorzellen. Verh Dtsch Ges Inn Med

Marangos PJ, Gazdar AJ, Carney DN (1982) Neuron specific enolase in human small-cell carcinoma cultures. Cancer Lett 15: 67–71

Moody TW, Pert CB, Gazdar AF et al. (1981) High levels of intracellular bombesin characterize human small-cell lung carcinoma. Science 214: 1246–1248

Pettengill OS, Sorenson GD, Wurster-Hill DH et al. (1980) Isolation and growth characteristics of continuous cell lines from small-cell carcinoma of the lung. Cancer 45: 906–918

Ratcliffe JG, Podmore J, Stack BHR, Spilg WGS, Gropp C (1982) Circulating ACTH and related peptide in lung cancer. Br J Cancer 45: 230

Sorenson GD, Pettengill OS, Brinck-Johnsen T et al. (1981) Hormone production by cultures of small-cell carcinoma of the lung. Cancer 47: 1289–1296

Whang-Peng J, Kao Shan CS, Lee EC et al. (1982) A specific chromosomal defect associated with human small cell lung cancer: Deletion 3p (14-23). Science 215: 181–182

Wolfsen AR, Odell WD (1979) Pro ACTH: use for early detection of lung cancer. Am J Med 66: 765

Peptide Hormone Production
Associated with Small Cell Lung Cancer

K. Havemann, W. Luster, C. Gropp, and R. Holle

Abteilung Hämatologie/Onkologie, Zentrum für Innere Medizin der Universität Marburg, Baldingerstrasse, 3550 Marburg, Federal Republic of Germany

Introduction

Small cell lung cancer (SCLC) is frequently associated with polypeptide hormone production, and a number of paraneoplastic syndromes have been reported. The neoplasia is thought to originate in the so-called Kulchitsky or K cells, small granular basal cells with endocrine properties found in the tracheobronchial mucosa (Tischler 1978). These K cells and the related SCLC and carcinoid tumor cells share APUD cell characteristics, such as cytoplasmic and membrane-bound dense core ("neurosecretory") granules, amine precursor uptake and decarboxylation of precursors to biogenic amines by L-dopa decarboxylase, and storage of amines and polypeptide products in the neurosecretory granules (Pearse 1969). It has therefore been suggested that APUD cells and their tumors have a neuroepithelial origin different from the endodermal origin of the rest of the respiratory mucosa and its related neoplasia, such as squamous cell carcinoma and adenocarcinoma (Pearse 1969). However, morphological findings of foci of squamous carcinoma and adenocarcinoma in SCLC (Carney et al. 1982), the simultaneous appearance of histologically different multiple lung cancer (Gazdar et al. 1981), and ultrastructural evidence that SCLC cells may undergo squamous cell metaplasia (Mackay et al. 1977) favor the hypothesis that all bronchial mucosa cells and the tumors arising from them have a common origin. Although polypeptide hormone production is more frequently associated with SCLC, other types of lung cancer with non-APUD cell characteristics frequently show hormone secretion (Rose 1979). Furthermore, hormonal peptides are produced by a number of other solid tumors, by lymphomas, and by different types of leukemia (Pflüger et al. 1981, 1982).

According to Roth (Roth et al. 1982), hormonal polypeptides and neurotransmitters occur very early in the evolution of cells, namely at the unicellular stage. They are probably involved in intercellular communication as local tissue factors. By the time the level of the multicellular organism has been reached, the relation between the secretory cell and the target cell has evolved extensively and become very diverse. However, the fundamental biochemistry of the system by which an agent carries a message from the secretory cell to the target cell remains ancient and highly conserved (Roth et al. 1982). Therefore, hormone production by human tumors may recall the early evolutionary stages of the cell development and may be a "universal concomitant of neoplasia" (Odell et al. 1977). Hormone production by SCLC is therefore only one example of this association.

Recent Results in Cancer Research. Vol. 97
© Springer-Verlag Berlin · Heidelberg 1985

Hormone Production by SCLC

Ectopic production of several peptide hormones has been described in patients with lung cancer during recent years. Peptide hormones which are frequently elevated in sera of patients with SCLC (Table 1) include adrenocorticotropic hormone (ACTH), melano-cyte-stimulating hormone (MSH), lipotropin (LPH), and β-endorphin (Gropp et al. 1981; Hansen et al. 1980; Krauss et al. 1981; Mackay et al. 1977; Odell et al. 1979; Ratcliffe et al. 1982). All these peptides arise from the common precursor molecule pro-opiomelanocortin by means of intra- or extracellular proteolytic cleavage (Marx 1983). The hypothalamic neurophysins antidiuretic hormone and oxytocin are also often elevated in blood specimens of patients with this tumor (Greco et al. 1981; Gropp et al. 1982; Hansen et al. 1980; Northetal 1980). The hormone most frequently exhibiting raised levels in SCLC is calcitonin (CT) (Greco et al. 1981; Hansen et al. 1980; Krauss et al. 1981; Luster et al. 1982; Mackay et al. 1977), which physiologically is a calcium-regulating peptide of the thyroid C cells. Parathormone is less often elevated in this histologic tumor type (Mackay et al. 1977). Similar findings have been reported for chorionic gonadotropin (Mackay et al. 1977). With the exception of gastrin, the other gastrointestinal peptides glucagon, secretin,

Table 1. Pertide hormones and NSE in serum or plasma of untereated patients with SCLC

	Patients (n)	Incidence (%)	Author
ACTH	75	29	Hansen et al. 1980
	50	30	Gropp et al. 1980
	68	38	Krauss et al. 1981
	63	24	Ratcliffe et al. 1982
α-MSH	43	19	Gropp et al. 1981
β-Endorphin	58	45	Gropp et al. 1981
LPH	24	54	Odell et al. 1979
ADH	41	39	Hansen et al. 1980
	61	48	North et al. 1980
	54	17	Greco et al. 1981
	66	30	Gropp et al. 1982
Oxytocin	61	30	North et al. 1980
CT	75	64	Hansen et al. 1980
	54	48	Gropp et al. 1980
	49	73	Krauss et al. 1981
	54	40	Greco et al. 1981
	135	56	Luster et al. 1982
PTH	43	27	Gropp et al. 1980
β-HCG	39	33	Gropp et al. 1980
Gastrin	69	20	Hansen et al. 1980
Glucagon	46	11	Hansen et al. 1980
Secretin, insulin VIP	46–65	≤5	Hansen et al. 1980
NSE	94	69	Carney et al. 1982

and insulin and the vasointestinal polypeptide VIP are only marginally elevated in sera of patients with SCLC (Hansen et al. 1980). Although the neuropeptide bombesin, known to induce gastrointestinal hormone secretion, is often present in cell extracts or supernatants of SCLC lines (Moody et al. 1981), its serum levels are low. This may be due to its high susceptibility to extracellular proteolytic cleavage, a phenomenon which it shares with the other gastrointestinal peptides. In contrast, neuron-specific enolase (NSE), a neuronal form of the glycolytic enzyme enolase present in brain, neuroendocrine tissue, and in tumors such as SCLC, is much more stable and shows the highest levels in plasma of patients with SCLC at diagnosis (Carney et al. 1982).

The percentage of elevated peptide hormone levels in plasma or sera of patients with SCLC at *diagnosis* (i.e., before start of treatment) is summarized in Table 1. Only studies performed on a large number of patients are included. The incidence of raised levels of a given peptide hormone varies according to the assay system, the antibodies used, and the upper limits employed in the different studies. As demonstrated in Table 1, in up to 70% of the patients, many of the hormones can be detected at diagnosis, and often several hormones are elevated at the same time. However, many of these blood levels are only marginally increased, and evidence of excessive production of these hormones may only be present in up to 25% of the patients (Carney et al. 1983).

None of these hormonal markers shows *specifity for SCLC,* since they are also elevated in sera of patients with other types of lung cancer. The only exception may be CT, which shows a rather low incidence in squamous cell and large cell carcinoma but exhibits comparatively high levels in adenocarcinoma (Luster et al. 1982; Mackay et al. 1977). A high specifity for SCLC has been claimed for NSE (Carney et al. 1982), results which have yet to be confirmed by other groups. Some of the hormones (ACTH, CT) are also raised in smokers and patients with chronic bronchitis and, in the case of CT, in patients with decreased kidney function. Marginally elevated values must therefore be interpreted with caution.

Peptide Hormones as Tumor Markers

Since SCLC is associated with the production of a number of peptide hormones, serum levels of these hormones were evaluated as markers for disease extent and response to therapy. Accordingly, ACTH, CT, neurophysins, and other tumor markers such as carcinoembryonic antigen (CEA) were measured in patients at diagnosis and sequentially following the start of therapy (Hansen et al. 1980; Krauss et al. 1981; Mackay et al. 1977; Ratcliffe et al. 1982). The serum levels were correlated with disease extent and the observed clinical response. While in some studies, the presence of increased levels of these markers showed a good correlation with disease extent, no such correlation was observed in others (Carney et al. 1983). In addition, some studies showed a close correlation between tumor response to cytotoxic therapy and decrease in hormone levels (Carney et al. 1983), suggesting that it is possible to monitor treatment and subsequently to change therapy according to the serum peptide hormone levels. All these data, however, have been obtained in rather small patient groups, and all were retrospective investigations.

The Tumor Markers CT, ACTH, and CEA in a Prospective Trial

In a multicenter trial for the treatment of SCLC by chemotherapy and radiation, the tumor markers CT, ACTH, and CEA were determined at the beginning of each cycle (eight cycles of chemotherapy followed by radiation of the primary tumor) and every 4 weeks after the follow-up period (Drings et al. 1983). The serum specimens were sent deep-frozen to the central laboratory. In spite of the organizational problems, it was possible to test more than 80% of the scheduled samples for the three markers mentioned.

Before Therapy

In the first 151 patients, we observed elevated (CT > 100 pg/ml, ACTH > 80 pg/ml, CEA > 5 ng/ml) or clearly pathological (CT > 200 pg/ml, ACTH > 150 pg/ml, CEA > 20 ng/ml) levels in 41% of the cases for CT, 15% for ACTH, and 49% for CEA. The incidence of the different marker combinations is demonstrated in Fig. 1. This incidence is lower than in retrospective studies, either due to the prospective design of the trial or due to problems arising from the transport of the samples to the central laboratory.
Comparing limited disease with extensive disease, only slight, statistically insignificant differences in the incidence of the markers could be observed at diagnosis. However, significantly higher marker levels were found in patients with distant metastasès as compared to patients without distant metastases, regardless of disease extent (Fig. 2a−c). On the other hand, a comparison of limited disease and extensive disease without distant metastasis showed no difference. These findings indicate that there is a relation between marker levels and tumor mass which in general is greater in patients with distant metastases. These data are in accordance with the results showing that the prognosis of limited disease and extensive disease without distant metastases is comparable (Ihde and Hansen 1981).
Further analyses show a correlation of increased CT and liver, bone, bone marrow, and brain metastases, while an increased level of CEA is only correlated with liver and bone

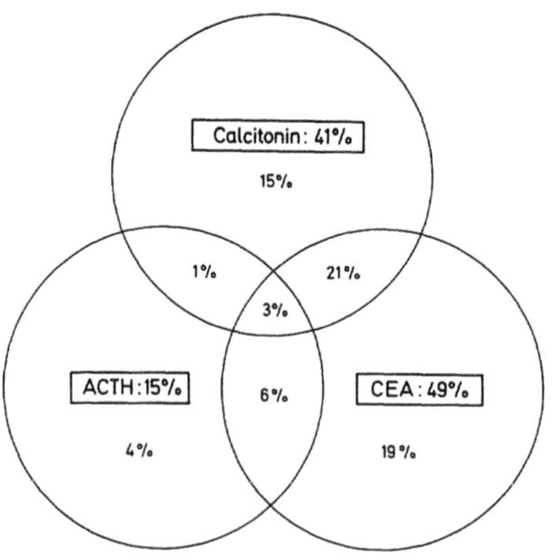

Fig. 1. Rate of elevated tumor marker levels before therapy (n = 151)

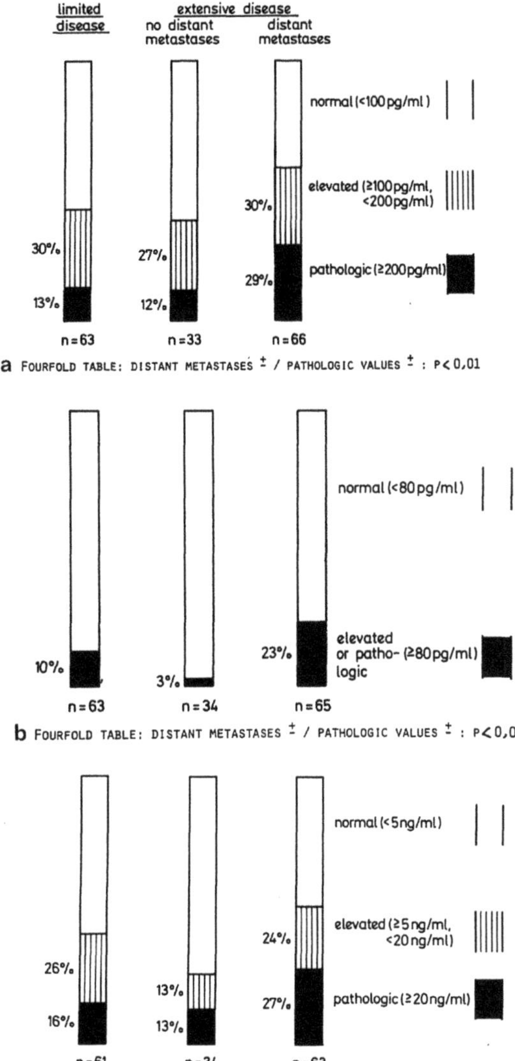

a FOURFOLD TABLE: DISTANT METASTASES ± / PATHOLOGIC VALUES ± : P < 0,01

b FOURFOLD TABLE: DISTANT METASTASES ± / PATHOLOGIC VALUES ± : P < 0,01

c FOURFOLD TABLE: DISTANT METASTASES ± / PATHOLOGIC VALUES ± : P = 0,07

Fig. 2a–c. Levels of tumor markers before therapy for **a** Calcitonin, **b** ACTH, and **c** CEA

metastases. However, in the individual patient, no correlation among the three tumor markers could be demonstrated.

During Therapy

In Table 2, the change in the tumor markers CT and CEA in patients with elevated or pathological values at diagnosis is compared with the quality of the therapeutic response (no response, minimal or partial response, and complete response). It is evident that a normalization of the increased levels mainly occurred in patients with complete remission or with partial and minimal response, whereas in general no decrease was observed in patients with no response. The few exceptions, for instance, no decrease of CEA in three

                                                    K. Havemann et al.

Table 2. Analysis of patients with initially elevated or pathological values of CT, ACTH, and CEA

CT level	Percentage of total	Clinical response		
		No response (*n*)	Minimal response partial response (*n*)	Complete response (*n*)
No decrease	21	11	0	0
Decrease (at least 20%), but not normal	21	1	9	1
Decrease to normal	58	1	18	12
ACTH level				
No decrease	44	1	2	1
Decrease (at least 20%), but not normal	0	0	0	0
Decrease to normal	56	0	2	3
CEA level				
No decrease	29	10	1	2
Decrease (at least 20%), but not normal	36	0	15	1
Decrease to normal	36	0	7	9

Chi squatre test: CT $P < 0.001$; ACTH $P > 0.05$; CEA $P < 0.001$

Table 3. Tumor marker during

The first cyle	Chest X-ray after the first cycle	
	Decrease in tumor size	No decrease in tumor size
CEA		
Decrease	*20*	1
No decrease	7	*8*
CT		
Decrease	*25*	1
No decrease	3	*4*

Italicized numbers emphasze positive correlation

cases of complete remission, were only seen in patients with primary elevated levels but never in those with true pathological values.

In most cases, the tumor marker levels reached normal values after one or two cycles of chemotherapy. Nearly all patients who reached normal tumor marker levels did so during the first four cycles.

Table 3 shows the correlation between the change in the tumor markers during the first cycle and the response of the primary tumor as indicated by chest radiographs. Only

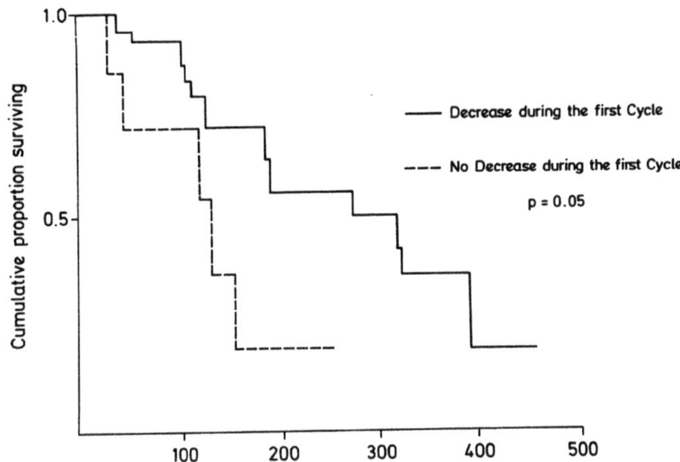

Fig. 3. Life table analysis (Kaplan Meier method) for patients with elevated calcitonin levels before therapy

patients with elevated or pathological markers are included in this table. There is a good correlation between the response of the primary tumor and the tumor markers. Discrepancies between the chest X-ray and the tumor markers may in some cases be due to metastases which exhibited a different response to therapy than the primary tumor. Patients whose response during the first cycle of chemotherapy was a decrease in the tumor marker had a significantly longer survival time than patients whose pathological values did not decrease. Median survival times for CT were 274 and 133 days and for CEA 307 and 198 days, respectively, while the overall median survival time for the trial was 300 days. The life table analysis for patients who responded by showing a decrease in CT and those who did not respond during the first cycle of chemotherapy is given in Fig. 3. The difference is statistically significant.

During Relapse

The rate of elevated tumor markers at the time of relapse is given in Fig. 4. The numbers are comparable to the values before therapy. In general, there was an increase for the same markers which were originally elevated at the time of diagnosis. Interestingly, in a number of cases, tumor markers which were initially normal became elevated during progressive disease (Table 4). In the example of a relapse given in Fig. 5, all three markers increase proportionately to one another. In other examples, one marker increased whereas the other markers which were originally elevated stayed normal. This was particularly true for CT. These data may indicate the overgrowth of a resistant marker-producing tumor cell clone.

The important question of whether an early increase in the marker level indicates a relapse even months before clinical evidence becomes apparent cannot yet be answered. It appears that this is only true for a limited number of patients who rapidly developed clearly pathological values.

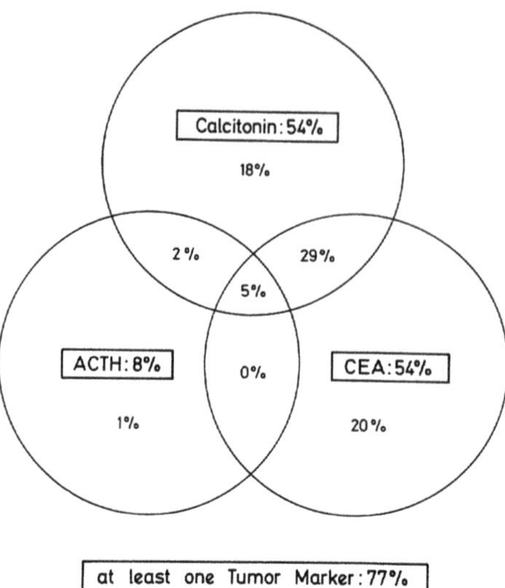

Fig. 4. Rate of elevated tumor marker levels at the time of relapse ($\pm$ 4 weeks) ($n = 74$)

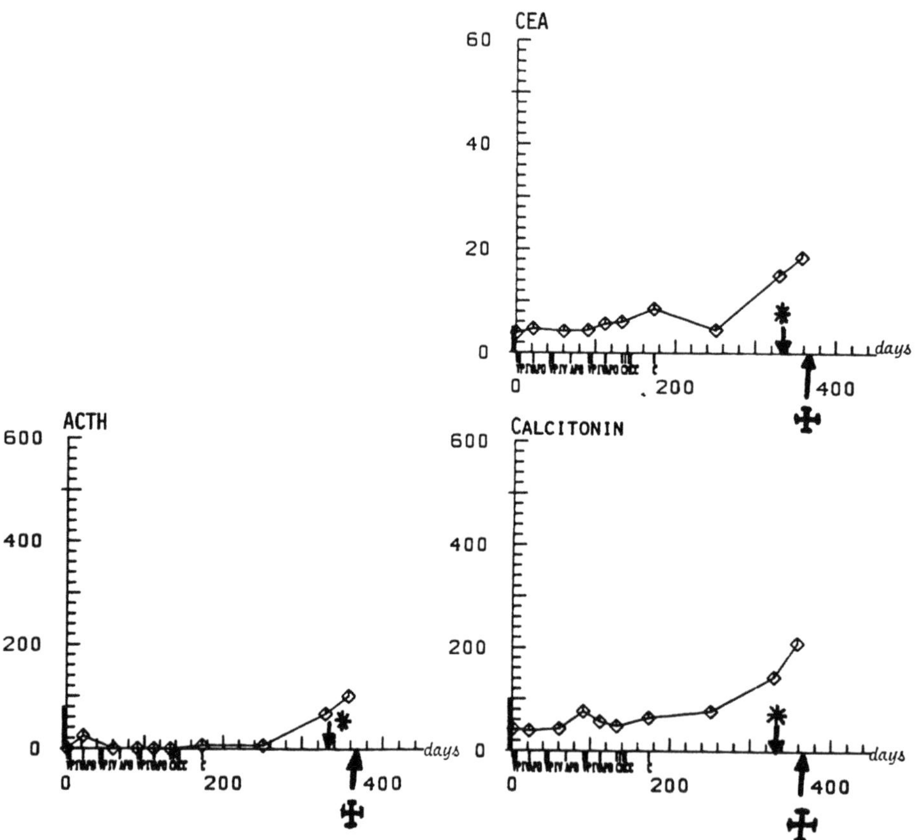

Fig. 5. Patient with extensive disease (partial remission) showing relapse on day 335 (liver and primary tumor)

Table 4. Analysis of initially normal tumor marker values

	CEA ($n = 59$)	calcitonin ($n = 51$)	ACTH ($n = 85$)
Always normal	83%	65%	92%
Initially normal, later elevated	17%	29%	7%
Initially normal, later pathologial	0%	6%	1%

Molecular Heterogeneity and Biological Importance of Peptide Hormones in SCLC

Pathologically elevated serum levels of immunologically determined peptide hormones are found in up to 70% of patients with SCLC (Havemann and Gropp 1979). On the other hand, a rather low incidence of paraneoplastic syndromes (about 5%) has been reported in these patients (Havemann and Gropp 1979; Roos et al. 1974). This discrepancy has led to speculation that these peptide hormones are biologically inactive and different from physiological hormones. Such molecular differences could be of biological importance and may also have diagnostic significance. For this reason, detailed investigations were carried out to characterize the peptide hormones of SCLC patients.

It has been demonstrated that tumors with an endocrine origin synthesize high molecular weight prohormones. For instance, in cell cultures C cell carcinomas were shown to produce high molecular weight calcitonins with a glycoprotein component (Jacobs et al. 1981), a finding which was confirmed by cell-free mRNA translation systems. It must be stressed, however, that such examinations led to results showing a wide range of molecular rates and molecular structures because of the use of different methods. Molecular weights of 8,000−55,000 daltons are described for CT prohormones (Desplan et al. 1980; Jacobs et al. 1981; Lips et al. 1978). McIntire et al. (1982) succeeded in sequencing a 1,000 dalton DNA coding a CT prohormone. This prohormone includes the amino acid sequence of physiological CT and contains another peptide which also has an effect on calcium metabolism (Craig et al. 1982).

From serum and tumor tissue of patients with SCLC, a stable 17,000 dalton core protein without a glycoprotein component and with an isoelectric point between 5.5 and 6 was characterized by gel infiltration and SDS-electrophoresis (Luster et al. 1982). It was shown by in vitro experiments with cell cultures that not only SCLC cells but also cells of adenocarcinoma and squamous and large cell lung carcinoma are able to secrete these prohormones, (Luster et al. 1983).

In pituitary gland tissue and in tumors of the hypophysis, results for ACTH prohormones are similar to those for CT in C cell tumors (Crine et al. 1978; Mains and Eipper 1977; Mains et al. 1977). The mRNA of a 31,000-dalton corticotropin-β-lipotropin prohormone was identified in bovine as well as human tissue. This RNA includes the code of ACTH and β-lipotropin and the sequence α-MSH, β-MSH, β-endorphin, α-MSH, β-lipotropin, and corticotropin like intermediate lobe peptide (Nakanishi et al. 1979; Takahashi et al. 1981).

In serum and tumor tissue of SCLC patients, ACTH-immunoreactive proteins with molecular weights as described for the pituitary gland prohormones were identified by gel filtration (Tanaka et al. 1978) and by affinity chromatography followed by gel filtration (Gropp et al. 1983). The reaction of these high molecular weight ACTH-immunoreactive proteins with antibodies to β-lipotropin was demonstrated in cell culture experiments

(Gropp et al. 1983). The presence of high molecular weight ACTH-β-lipotropin immunoreactive precursor molecules may explain the appearance of ACTH in circulating immune complexes of patients with SCLC, since the ACTH-β-lipotropin prohormone, which is normally not present in circulation, may have the function of an autoantigen (Havemann 1979).

The biological importance of peptide hormones for tumor growth is demonstrated by results showing an increase in the tumor growth rate of lung tumor cells in vitro after the addition of peptide hormones to the culture medium (Luster et al. 1984). Some of these peptide hormones, such as ACTH, bombesin, and ADH, seem to stimulate preferentially the growth of tumor stem cells. It remains to be examined, however, to what extent the stimulation of in vitro tumor cell growth is also influenced by the addition of high molecular weight prohormones to the cell culture medium.

These findings led to the hypothesis that SCLC cells sustain their own growth by means of tumor-transforming factors (Sherwin and Todaro 1983) and peptides with hormonal activity according to an autocrine or paracrine regulation process (Todaro and Sporn 1980). Through the transforming process, the tumor cell may return to a phylogenetically very early stage, at which it is then responsible for the secretion of hormone precursor molecules. This hypothesis is supported by the identification of high molecular weight proteins with binding sites for antibodies against ACTH and β-endorphin in protozoans, fungi, and bacteria (Roth et al. 1982). Another explanation for the secretion of precursor molecules by tumor cells could be the deficiency of enzymes necessary for the processing of the hormones. Moreover, a protein kinase coded by an oncogene may inactivate the enzmyes participating in the processing directly by phosphorylation or indirectly by phosphorylation of one of the enzymes necessary for the biosynthesis of the processing enzymes.

Recent results indicate that cellular heterogeneity of SCLC tumors expresses itself in the peptide hormone biosynthesis by these tumors. Immunohistochemical experiments show that only part of the tumor cells are able to synthesize and accumulate hormone immunoreactive proteins (Gropp et al. 1981). This heterogeneity of solid tumors has been also detected by cytophotometric methods (Haskill et al. 1983). Thus, Vindelov et al. (1980) found two or more different cell clones in 21% of the examined lung tumors. This heterogeneity probably results from the selection process in the tumor tissue and is promoted by the genetic instability of the tumor cell. It is tempting to speculate that the peptide hormone-producing clones maintain growth of other subclones in a way equivalent to a paracrine regulation process.

Conclusion

SCLC is associated with a number of peptide hormones which are produced by the tumor cells themselves. In combination with other diagnostic procedures, these hormones can be used as tumor markers for staging and monitoring therapy. Peptide hormone determination seems to be of little value for early prediction of tumor relapse.

Peptide hormones associated with SCLC are high molecular weight prohormones. This has been shown for ACTH and CT. These hormones may be biologically important for the tumor growth of SCLC according to an autocrine or paracrine process.

References

Brereton HD, Matthews MJ, Costa M (1978) Mixed anaplastic small cell and squamous cell carcinoma of the lung. Ann Intern Med 88: 805–806

Carney DN, Marangos PJ, Ihde DC (1982) Serum neuron-specific enolase: a marker of disease extent and response to therapy in patients with small cell lung cancer. Lancet 1: 583–585

Carney DN, Broder L, Edelstein M, Gazdar AF, Hansen M, Havemann K, Matthews MJ, Sorenson GD, Vindelov L (1983) Experimental studies of the biology of human small cell lung cancer. Cancer Treat Rep 67: 27–35

Craig RK, Hall L, Edbrooke MR, Allison J, MacIntyre I (1982) Partial nucleotide sequence of human calcitonin precursor mRNA identifies flanking cryptic peptides. Nature 295: 345–347

Crine P, Ianoulakis C, Seidah NG, Gossard F, Pezalla PD, Chretien M (1978) Biosynthesis of β-endorphin from β-lipotropin and a larger molecular weight precursor in rat pars intermedia. Proc Natl Acad Sci USA 75: 4719–4722

Desplan C, Benicourt C, Jullienne A (1980) Cell free translation of mRNA coding for human and murine calcitonin. FEBS Lett 177: 89–92

Drings P, Holle R, Havemann K, Harms V (1983) The tumor markers CEA, calcitonin, and ACTH in a prospective trial of small cell lung cancer. In: Spitzy KH, Karrer K (eds) Proceedings 13th Int Congress Chemother, Vienna, 274: pp 45–48

Gazdar AF, Carney DN, Guccion JG, Baylin SB (1981) Cellular origin and relationship to other lung tumors. In: Greco FA, Oldham RK, Bunn PA (eds) Small cell lung cancer. Grune and Stratton, New York, pp 145–176

Greco AF, Hainsworth J, Sismann A (1981) Hormone production and paraneoplastic syndromes. In: Greco AF, Oldham RK, Bunn PA (eds) Small cell lung cancer. Grune and Stratton, New York, pp 177–224

Gropp C, Luster W, Havemann K, Lehmann FG (1981) ACTH, calcitonin, α-MSH, β-endorphin, parathormone and β-HCG in sera of patients with lung cancer. In: Uhlenbruck G, Wintzer G (eds) CEA und andere Tumormarker. Tumor Diagnostik, Leonberg, pp 358–363

Gropp C, Havemann K, Kalbfleisch H, Luster W, Sostmann H (1982) Antidiuretisches Hormon bei Patienten mit Bronchialkarzinom. Dtsch Med Wochenschr 107: 977–980

Gropp C, Luster W, Havemann K (1983) High molecular adrenocorticotropin and calcitonin immunoreactive proteins as biosynthesis products of small cell lung tumors. Acta Endocrinol [Suppl] (Copenh) 253: 16–17

Hansen M, Hammer M, Hummer L (1980) ACTH, ADH, and calcitonin concentrations as marker of response and relapse in small cell carcinoma of the lung. Cancer 46: 2062–2067

Haskill S, Kivinen S, Nelson K, Fowler WC Jr (1983) Detection of intratumor heterogeneity by simultaneous multiparameter flow cytometric analysis with enzyme and DNA markers. Cancer Res 43: 1003–1009

Havemann K, Gropp C (1979) Biological and immunological aspects of small cell carcinoma of the lung in relation to ectopic hormone production. Biomedicine 30: 186–194

Havemann K, Gropp C, Scheuer A, Schärfe T, Gramse M (1979) ACTH-like activity in immune complexes of patients with oat-cell carcinoma of the lung. Br J Cancer 39: 43–50

Ihde DC, Hansen HH (1981) Staging procedures and prognostic factors in small cell carcinoma of the lung. In: Greco FA, Oldham RK, Bunn PA (eds) Small cell lung cancer. Grune and Stratton, New York, pp 261–284

Jacobs JW, Lund PK, Potts JT, Bell NH Jr, Habener JF (1981) Procalcitonin is a glycoprotein. J Biol Chem 256: 2803–2807

Krauss S, Macy S, Ichiki AT (1981) A study of immunoreactive calcitonin, ACTH and CEA in lung cancer and other malignancies. Cancer 47: 2485–2492

Lips CJM, Van Der Sluys Veer J, Van Der Donk JA, Van Dam RH, Hackeng WHL (1978) Common precursor molecule as origin for the ectopic hormone-producing tumor syndrome. Lancet 1: 16–18

Luster W, Gropp C, Sostmann H, Kalbfleisch H, Havemann K (1982) Demonstration of immunoreactive calcitonin in sera and tissues of lung cancer patients. Eur J Cancer Clin Oncol 18: 1275–1283

Luster W, Gropp C, Havemann K (1983) Peptide hormone-synthesizing lung tumor cell lines: establishment and first characterization of biosynthetic products. Acta Endocrinol [Suppl] (Copenh) 102: 24–25

Luster W, Gropp C, Loeck MR, Havemann K (1984) Modification of tumor cell proliferation and peptide hormone secretion. In: Protides of the biological fluid. Pergamon (in press)

MacIntyre I, Hillyard CJ, Murphy PK, Reynolds JJ, Gaines RE, Craig RK (1982) A second plasma calcium-lowering peptide from the human calcitonin precursor. Nature 300: 460–462

Mackay B, Osborne BM, Wilson RA (1977) Ultrastructure of lung neoplasma. In: Strauss MJ (ed) Lung cancer – clinical diagnosis and treatment. Grune and Stratton, New York, pp 71–84

Mains RE, Eipper BA (1977) Coordinate synthesis of corticotropins and endorphins by mouse pituitary tumor cells. J Biol Chem 253: 651–655

Mains RE, Eipper BA, Ling N (1977) Common precursor to corticotropins and endorphins. Proc Natl Acad Sci USA 74: 3014–3018

Marx JL (1983) Synthesizing the opioid peptides. Science 220: 395–397

Moody TW, Pert CB, Gazdar AF (1981) High levels of intracellular bombesin characterize human small cell lung carcinoma. Science 214: 1246–1248

Nakanishi S, Inoue A, Kita T, Nakamura M, Chany ACY, Cohen SN, Numa S (1979) Nucleotide sequence of cloned cDNA for bovine corticotropin-β-lipotropin precursor. Nature 278: 423–427

North WG, Maurer H, Valtin H, O'Donell JF (1980) Human neurophysins as potential tumor markers for small cell carcinoma of the lung: Application of specific radioimmunoassays. J Clin Endocrinol Metab 51: 892–897

Odell WD, Wolfsen AF, Yoshimoto Y (1977) Ectopic peptide synthesis: a universal concomitant of neoplasia. Trans Assoc Am Physicians 40: 204–225

Odell WD, Wolfsen AR, Bachelot I, Hirose FM (1979) Ectopic production of lipotropin by cancer. Am J Med 66: 631–638

Pearse AGE (1969) The cytochemistry and ultrastructure of polypeptide hormone-producing cells of the APUD series and the embryologic, physiologic and pathologic implications of the concept. J Histochem Cytochem 17: 303

Pflüger KH, Gramse M, Gropp C, Havemann K (1981) Ectopic ACTH production with autoantibody formation in a patient with acute myeloblastic leukemia. N Engl J Med 305: 1632–1636

Pflüger KH, Gropp C, Havemann K (1982) Ectopically produced calcitonin in human hemoblastoses. Klin Wochenschr 60: 667–672

Ratcliffe JG, Podmore J, Stack BHR, Spilg WGS, Gropp C (1982) Circulating ACTH and related peptides in lung cancer. Br J Cancer 45: 230–238

Roos BA, Okano K, Deftos LJ (1974) Evidence for a procalcitonin. Biochem Biophys Res Commun 60: 1134–1140

Rose DP (1979) The ectopic production of hormones by tumors. In: Rose DP (ed) Endocrinology of cancer II. CRC Press, Boca Raton, pp 95–124

Roth J, LeRoith D, Shiloach J (1982) The evolutionary origin by hormones, neurotransmitters and other extracellular chemical messengers. N Engl J Med 306: 523–527

Sherwin SA, Todaro GJ (1983) Transforming growth factors and human lung cancer. In: Greco FA (ed) Biology and management of human lung cancer. Martinus Nighoff, Boston, pp 37–50

Takahasi H, Teranishi Y, Nakanishi S, Numa S (1981) Isolation and structural organization of the human corticotropin-β-lipotropin precursor gene. FEBS Lett 135: 97–102

Tanaka K, Nicholson WE, Orth DN (1978) The nature of the immunoreactive lipotropins in human plasma and tissue extracts. J Clin Invest 62: 94–104

Tischler AS (1978) Small cell carcinoma of the lung, cellular origin and relationship to other neoplasms. Semin Oncol 5: 244

Todaro GJ, Sporn MB (1980) Autocrine secretion and malignant transformation of cells. N Engl J Med 303: 878–890

Vindelov L, Hansen HH, Christensen IJ, Spang-Thomasen M, Hirsch FR, Hansen M, Nissen NI (1980) Clonal heterogeneity of small cell anaplastic carcinoma of the lung demonstrated by flow-cytometric DNA analysis. Cancer Res 40: 4295–4300

Development of Three Human Small Cell Lung Cancer Models in Nude Mice

H. H. Fiebig, H. A. Neumann, H. Henß, H. Koch, D. Kaiser, and H. Arnold

Medizinische Klinik der Universität Freiburg, Abteilung für Hämatologie und Onkologie, Hugstetter Strasse 55, 7800 Freiburg, Federal Republic of Germany

Introduction

Most human tumors can be successfully grown in athymic nude mice (Sharkey 1978; Fiebig and Löhr 1984). Little experience has been reported on the transplantation of human small cell lung cancer (SCLC), because this tumor category is usually not treated surgically. In this paper we report on our results with transplantation of seven SCLC into nude mice. Three rapidly growing tumors were selected as tumor models. Their characteristics and responsiveness to drug treatment are described. Tumor response in the nude mouse and in the human patient is compared.

Materials and Methods

Animals

Athymic nude mice of NMRI genetic background were used at 4−6 weeks of age. They were bred in our own nude mouse colony. The animals were kept in macrolon cages set in laminar flow racks. They were maintained as described by Fortmeyer and Bastert (1981). Tumors from men were implanted into male nude mice and tumors from women into female animals.

Tumors

Seven histologically confirmed SCLC obtained by surgery were transplanted into nude mice. In five cases a lymph node metastasis and in two cases distant metastases (a skin and a chest wall metastasis) were implanted. Distant metastases were present in four patients and lymph node metastases in three patients. Tumor slices with diameters 5 mm × 5 mm × 0.5−1 mm were implanted subcutaneously into the flanks of the animals, usually sixteen fragments into four nude mice in the first passage. When the tumors reached diameters of 1−1.5 cm they were subpassaged and the remaining tumor material was studied histologically. The product of two perpendicular diameters was taken as a measure of tumor size. Relative tumor size values were calculated as follows: tumor size on day X divided by tumor size on day 0 at the time of randomization and multiplied with 100. The effect of treatment was classified as remission (product of the two diameters less than 50% of initial value), minimal regression (51%−75%), no change (76%−124%), or progression ($\geq$ 125% of initial value after 3−4 weeks).

Recent Results in Cancer Research. Vol. 97
© Springer-Verlag Berlin · Heidelberg 1985

Demonstration of Carcinoembryonic Antigen in the Tumor

The localization of carcinoembryonic antigen (CEA) in the tumor section was determined using the indirect immune peroxidase method (Wittekind et al. 1982).

Determination of Esterase-D and LDH

Human esterase-D and the homologous murine esterase-10 and also human and murine LDH were used to demonstrate the human origin of the tumors. They produce different bands in starch gel electrophoresis. The technique was applied as described by Bissbort (1981).

Chemotherapy in Vivo

Drugs were given in schedules derived from clinical use. Dosages and application are shown in Figs. 3 and 4. A dose about the LD_{20} after 28 days was used, which was considered as the maximum tolerable dose.

Tumor Stem Cell Assay In Vitro

The tumor cell cultures were performed as described elsewhere (Neumann et al. 1983). Briefly, tumors growing in nude mice were disaggregated mechanically to give single-cell suspensions. 10^5 tumor cells were seeded in Iscove's modification of Dulbecco's medium (Gibco) containing 5% fetal calf serum, and drug solution. Methylcellulose at a final concentration of 0.9% (w/v) was used as a viscous support. No conditioned medium was added. The tumor cell suspension was exposed continuously to drugs in various concentrations, as shown in Figs. 6 and 7. After 8–10 days the cultures were scored under an inverted microscope. Cell agregates of more than 40 cells were considered as colonies. The effect of the drugs was expressed as reduction in colony formation in comparison with untreated controls.

Results

Take Rate and Growth Behavior

Out of the total of 68 lung cancers of different histologies, 57 yielded viable tumor tissue in the nude mouse as proven by histological examination. A rapid tumor growth, defined as tumor size of at least 60 mm^2 (a × b) after 90 days, was observed in 40 tumors (59%), and a slow growth in 17 tumors (25%). Most of the rapidly growing tumors, and in rare cases also initially slow-growing tumors, were transferred to serial passages (Table 1). Forty lung cancers (59%) were subpassaged at least three times. The take rate and growth behavior of the seven SCLC were similar to those lung cancers with other histologies, although the number of transplanted SCLC was small.

The growth velocity of the SCLC was quite different in the first passage. After initial shrinkage within the first 2–3 weeks, three tumors showed rapid growth after 2 months, whereas two tumors grew progressively after lag phases of 4 and 6 months. One tumor

Table 1. Take rate and growth behavior of human lung cancers after transplantation into nude mice

Tumor histology	Total number	Tumor take[a]		Rapid growth[b]		Slow growth[c]		Serial passage[d]	
		n	%	n	%	n	%	n	%
Small cell	7	6	86	3	43	3	43	3	43
Epidermoid	35	30	86	24	69	6	17	25	71
Adeno	16	15	88	9	56	6	31	8	50
Large cell	6	5	83	3	50	2	33	3	50
Unclassified	2	1		1		0		1	
Total	68	57	84	40	59	17	25	40	59

[a] Histologic demonstration of viable tumor tissue
[b] Size (a × b) of at least one tumor $\geq 60\ mm^2$ in the first passage after 90 days
[c] Size (a × b) of at least one tumor $\leq 59\ mm^2$ in the first passage after 90 days
[d] Continuous growth of at least three passages

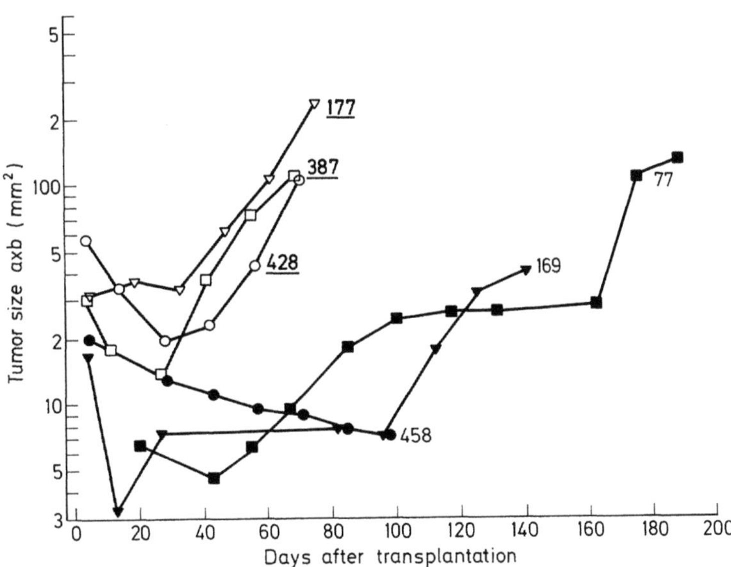

Fig. 1. Growth of human small cell lung cancers after transplantation into nude mice. Tumor size of the largest tumor in the first passage of histologically confirmed tumors. *Underlined* tumors were subpassaged

remained stationary; however, it was composed of viable tumor tissue, as demonstrated by histological examination. In Fig. 1 the growth of the largest tumor in the first passage is shown. The three tumors underlined were transferred to serial passage. The histological appearance of the mouse-grown tumors was strikingly similar to that of the original tumor in the patient (Fig. 2).

The tumors were circumscribed in the mouse but not encapsulated. Occasionally they infiltrated surrounding muscles. The histological appearances were typical of SCLC. Extensive areas of necrosis were frequently found in large tumors. In some cases tumor cells were present in veins draining the tumor. Metastases were not observed. Four of six

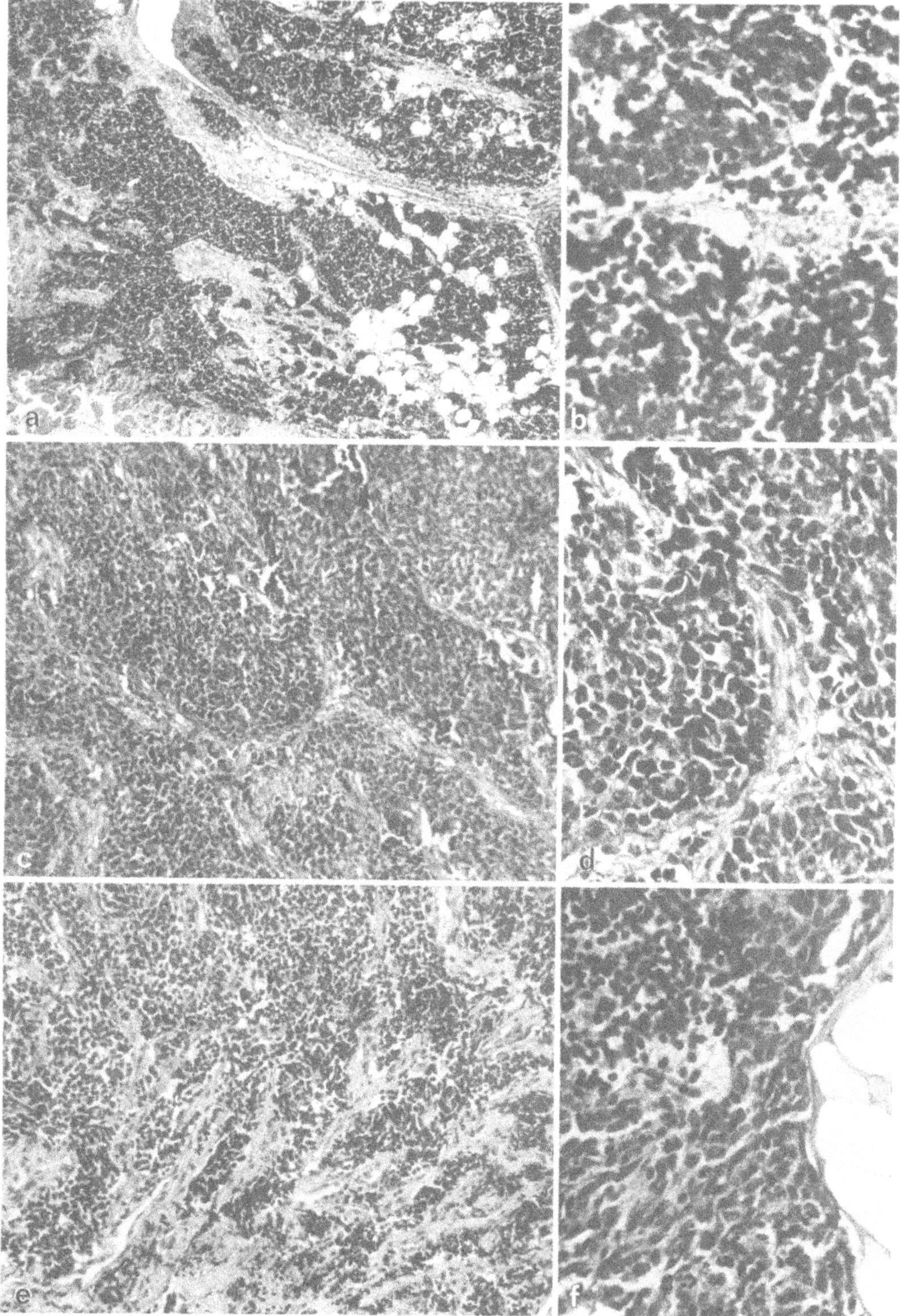

Fig. 2a—f. Histologic picture of human SCLC LXF 428 of the intermediate cell type. **a, b** Donor tumor; H & E, × 63, × 400, **c, d** 3rd passage; × 160, × 400, **e, f** 6th passage; × 160, × 400

Table 2. Properties of tumor models for human small cell cancer of the lung in nude mice

Tumor no. LXF	Histology SCLC	Tumor origin	Prior chemotherapy	Transplantation month/year	Tissue CEA
177	Intermed.	LN	6 × ACO	3/80	Positive
387	Intermed.	LN	None	8/82	Positive
428	Intermed.	Chest wall metastasis	6 × ACO	12/82	Negative

Tumor no. LXF	Take rate[a] (%)	Median doubling time[b] (days)	Time to reach 100 mg (days)	Serial passage no. (3/84)	Frozen in N2 from passage no.
177	63	9.8	30	20	6, 10, 20
387	86	10.7	34	7	3, 6
428	64	15.0	45	6	5

[a] Last 3−6 serial passages
[b] Single tumors of the last passages, $n = 10-22$
LXF, lung cancer xenograft Freiburg; *LN*, lymph node; *Intermed.*, intermediate cell type; *ACO*,
adriamycin + cyclophosphamide + vincristine; *N2*, liquid nitrogen

tumors growing in nude mice produced CEA as demonstrated in tumor sections. CEA positivity or negativity was retained in serial passages, being studied over 18 subpassages in one case.

Characterization of Three Tumor Models

Three SCLC growing regularly in serial passages were selected as tumor models. Properties of these tumors are described in Table 2. LXF 177 was derived from a lymph node metastasis of a patient who had received prior chemotherapy with the ACO combination. LXF 387 was not pretreated in the patient. LXF 428 originated from a chest wall metastasis of a patient who had previously received the ACO combination. The take rates in serial passages ranged from 63% to 86%, which was inferior to the take of other tumor categories (Fiebig et al., to be published). This may be due to the marked tendency of SCLC to become necrotic, so that in some instances fragments with no viable tissue can be transplanted. All three tumor models were frozen in liquid nitrogen. The human origin of the tumors was demonstrated by the isoenzyme analysis. The xenografts presented bands of human LDH and human esterase-D.

Responsiveness to Chemotherapy

The responsiveness to drugs was studied in subcutaneously growing tumors in nude mice and in tissue culture using the methodology of the stem cell assay. Figure 3 presents the response of LXF 177 to the combinations adriamycin + cyclophosphamide + vincristine and VP-16 + cisplatin, which effected remissions. Two cycles resulted in a response

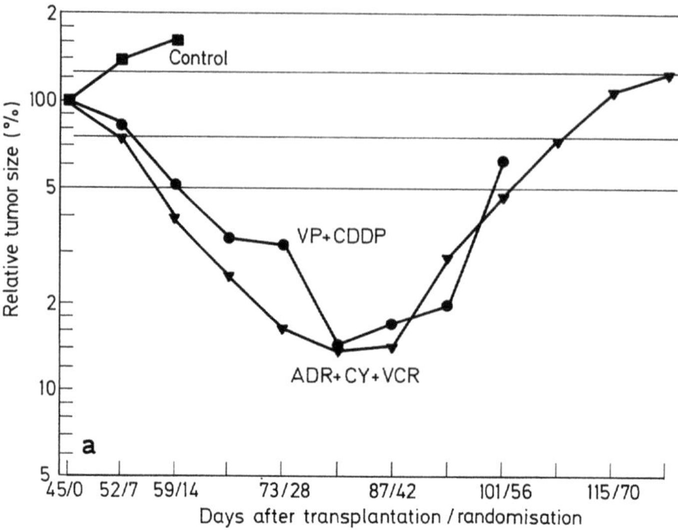

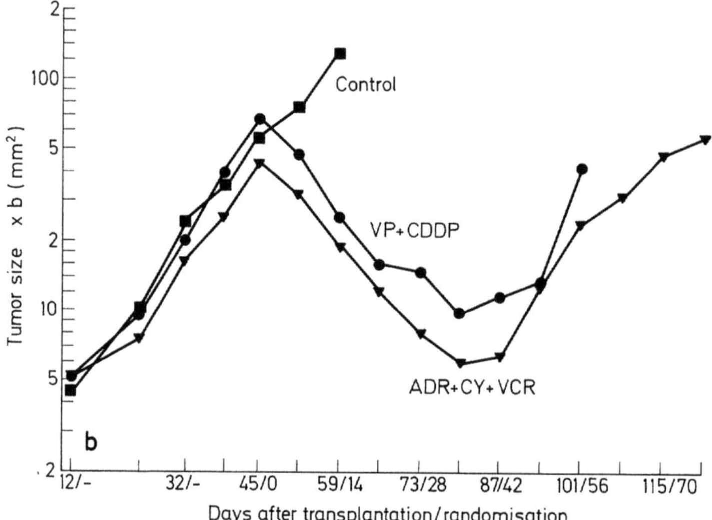

Fig. 3a, b. Responsiveness of SCLC 177 in nude mice after treatment with the combinations VP-16 (24 mg/kg, days 1–3, 15–17, SC) + cisplatin (3.2 mg/kg days 1–2, 15–16, SC); and adriamycin (4 mg/kg, days 1 + 15, IV) + cyclophosphamide (150 mg/kg, days 1 + 15, SC) + vincristine (0.5 mg/kg, day 1 + 15, IP). **a** Relative tumor size; **b** absolute tumor size (product of two diameters)

duration of 6 weeks. The xenograft LXF 387 responded in a similar way. The two patients received the two combinations, and also experienced partial remission with both. The survival times of the patients were 459 and 580 days, respectively. LXF 177 had been pretreated with six cycles of the ACO combination in the patient. Treatment was stopped in the patient so that no resistance had developed. The xenograft LXF 177 was still responding to this combination. The response of LXF 387 to single-agent therapy is shown in Fig. 4. High-dose cisplatin, vindesine, and adriamycin effected partial remissions, whereas VP-16 and cyclophosphamide were somewhat less active.

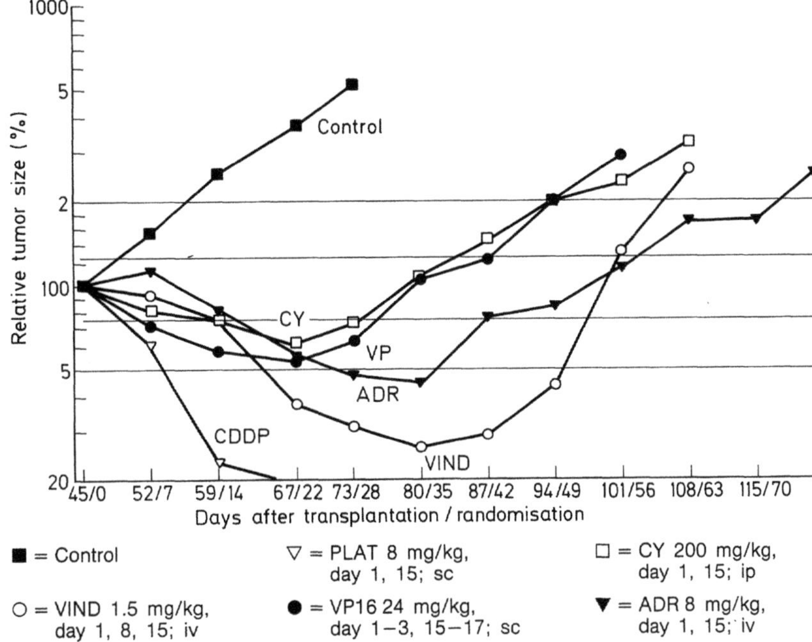

Fig. 4. Responsiveness of SCLC 387 to five drugs in nude mice

Drugs can also be tested in cell culture. In semisolid media such as agar or methylcellulose, single-cell suspensions developed to tumor colonies after 7–14 days (Fig. 5). An example of the effect of 8 drugs against LXF 177 is shown in Fig. 6. The dose appropriate to the clinical situation is a critical question. To determine appropriate dosages, studies comparing the effect in vivo in nude mice and in vitro in the stem cell assay are in progress. Furthermore, new drugs can also be tested in vitro. Figure 7 illustrates this with reference to N-methylformamide, 1,2,4-triglycidyl-urazol (TGU), and tiazofurine, three drugs that are currently in phase I clinical trial.

Discussion

Most human tumor categories can successfully be grown in nude mice (Fiebig and Löhr 1984). Limited experience has been reported with human SCLC, as this tumor is usually treated by chemo- and radiotherapy and not by surgery. Our take rate is somewhat higher than that published by Gazdar et al. (1981), who reported a take rate of 45% (13/29) after subcutaneous inoculation. Ten tumor lines were established. After intracerebral inoculation the take rate was 86%. With intracerebral heterotransplants 10–1,000 times fewer cells are required to induce a tumor than with subcutaneous inoculation (Gazdar et al. 1981). Similar findings were reported by Chambers et al. (1981), who used continuous cell lines of SCLC established in cell culture.
Human SCLC growing in serial passage in nude mice are useful models for the study of tumor biology and responsiveness to known and new drugs. As shown by histological and immunohistological examinations (CEA), the tumors growing in mice retained the characteristics of the original donor tumors.

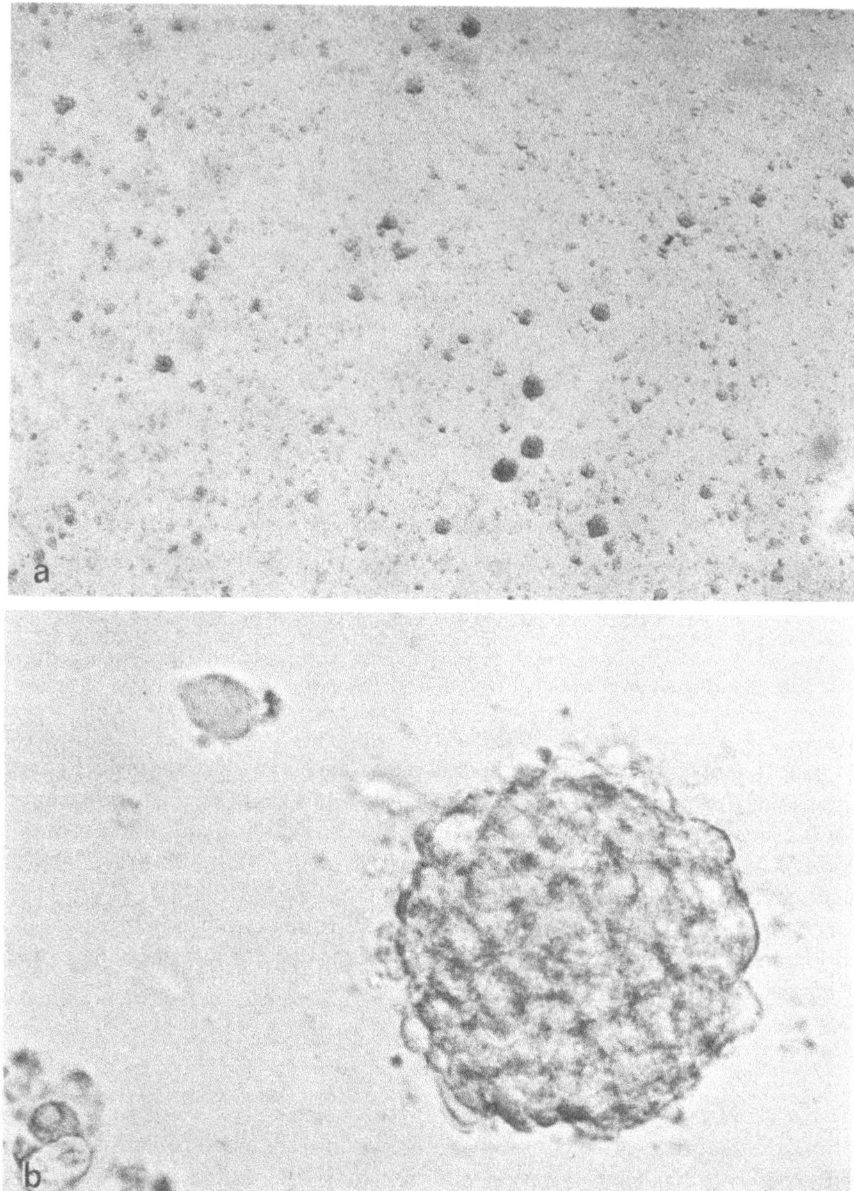

Fig. 5a, b. In vitro colony formation of human SCLC 387/6. **a** × 8; **b** × 128

For therapeutic studies an important question is whether the tumors respond in the same manner in the nude mouse system and in the patient. The four comparisons we made in two patients showed a remission in the xenograft and in the patient in all four cases. Identical results were reported by Shorthouse et al. (1980). Five SCLC responded to drug treatment, whereas one SCLC was resistant in the nude mouse and in the patient. Our overall experience in the comparison of tumor response includes 34 tumors of different origins, in which 50 comparisons were performed. Xenografts gave a correct prediction for resistance

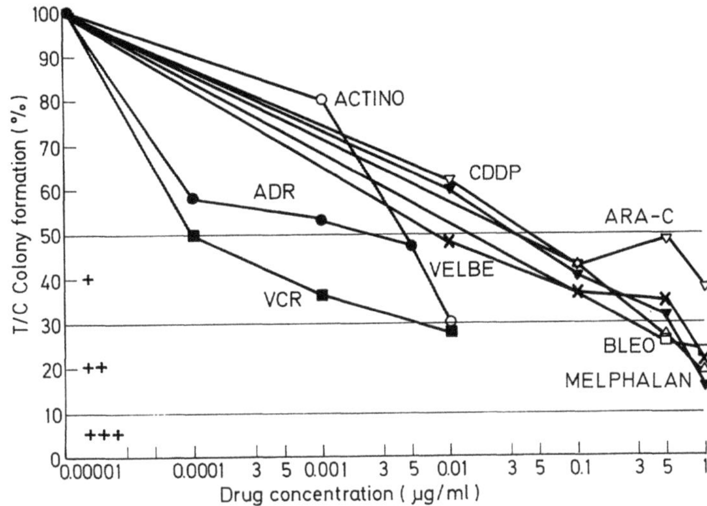

Fig. 6. In vitro sensitivity of SCLC 177. Inhibition of colony formation against eight drugs, expressed as percentages of the values in untreated controls

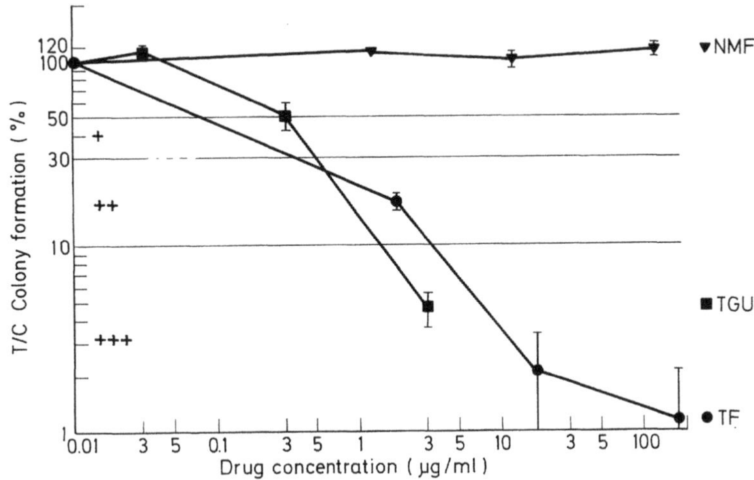

Fig. 7. In vitro sensitivity of SCLC 177 against *N*-methylformamide, TGU, and tiazofurin, expressed as percentages of control values

in 97% and for tumor response in 92% (Fiebig et al. 1984). These high rates of accurate prediction validate human tumor xenografts as tumor models for testing of new drugs and new combinations.

Summary

The transplantation of seven human small cell lung cancers (SCLC) into athymic nude mice resulted in the development of three tumor lines that are suitable for study of tumor biology and for tests of new drugs and combinations. They were characterized and the response to known drugs was determined. An identical tumor response was observed in the nude mouse system and in the patient in all four comparisons available.

Acknowledgements. This project was supported by grant PTB 8466 from the Bundesministerium für Forschung und Technologie [Federal Ministry of Research and Technology].
We thank Dr. J. Venditti, Drug Evaluation Branch, Developmental Therapeutic Program, Division of Cancer Treatment, National Cancer Institute, Bethesda, for supplying *N*-methylformamide and tiazofurine.

References

Bissbort S (1981) Zur Genetik der Phosphoglucomutasen des Menschen. Medical dissertation, University of Freiburg

Chambers WF, Pettengill OS, Sorenson GD (1981) Intracranial growth of pulmonary small cell carcinoma cells in nude athymic mice. Exp Cell Biol 49: 90–97

Fiebig HH, Löhr GW (1984) Wachstum menschlicher Malignome und Entwicklung von Tumormodellen in der thymusaplastischen Nacktmaus. Med Welt 35: 52–58, 81–86

Fiebig HH, Schuchhardt C, Henss H, Fiedler L, Löhr GW (1984) Comparison of tumor response in nude mice and in the patients. Behring Inst Mitt 74: 343–352

Fortmeyer HP, Bastert G (1981) Breeding and maintenance of nu/nu mice and rnu/rnu rats. In: Bastert G, Fortmeyer HP, Schmidt-Matthiesen H (eds) Thymus aplastic nude mice and rats in clinical oncology. Fischer, Stuttgart New York, pp 25–38

Gazdar AF, Carney DN, Sims HL, Simmons A (1981) Heterotransplantation of small-cell carcinoma of the lung into nude mice: comparison of intracranial and subcutaneous routes. Int J Cancer 28: 777–783

Neumann H, Engelhardt R, Löhr GW, Fauser AA (1983) Cloning of tumor cells from patients with melanoma, colon carcinoma, gastric tumor and small cell carcinoma of the lung in methylcellulose monolayer system. Verh Dtsch Krebsges 4: 155

Sharkey FE (1978) Histopathological observations on a nude mouse colony. In: Fogh J, Giovanella BC (eds) The nude mouse in experimental and clinical research. Academic, New York San Francisco London, pp 75–93

Shorthouse AJ, Peckham MJ, Smyth JF, Steel GG (1980) The therapeutic response of bronchial carcinoma xenografts: a direct patient-xenograft comparison. Br J Cancer [Suppl 4] 41: 142–145

Wittekind C, von Kleist S, Sandritter W (1982) Vergleichende immunhistologische Untersuchungen über die Lokalisation des carcinoembryonalen Antigens in gut- und bösartigem Mammagewebe. Acta Histochem [Suppl] (Jena) 25: 83–87

Diagnostic Procedures in Small Cell Lung Carcinoma

P. Drings, R. König, and I. Vogt-Moykopf

Krankenhaus Rohrbach, Klinik für Thoraxerkrankungen der LVA Baden,
Amalienstrasse 5, 6900 Heidelberg 1, Federal Republic of Germany

Introduction

Besides the histological type, the anatomical extent of the tumor and the general condition of the patient determine the prognosis and the therapeutic concept (operation, radiotherapy, or chemotherapy) in bronchial carcinoma (Carr 1973; Carter 1979; Mountain and Hermes 1979; Ihde and Hansen 1981; Hansen and Dombernovsky 1981). In addition, the degree of malignancy, the rate of tumor cell doubling, the demonstration of vascular infiltration, the presence of clinical symptoms of the tumor, and the age and sex of the patient are important (Edmonson et al. 1976, Maurer et al. 1980), as are any concomitant conditions. These various biological factors are examined in an extensive program before the beginning of therapy. The result of this program provides additional pointers to the prognosis and allows comparison of the results of different therapeutic studies. For this purpose, a standardized system of investigation is required.
The general condition of the patient can be evaluated in accordance with the performance scale of Karnofsky et al. (1948) (Table 1) or, as suggested by the American Joint Committe (1979); according to the Zubrod scale (Table 2).
For evaluation of the anatomical stage of tumor spreading there are also two suggestions. In the very few patients who are potentially operable the tumor stage is determined before treatment according to the TNM system of the UICC (1978) (Table 3). This classification is supplemented postoperatively with reference to the histopathological investigation of the resection preparation (p-TNM). As Mountain and Hermes (1979), using the criteria of the American Joint Committee for Cancer Staging and End Results Reporting (1979), were able to show, this allows significant prognostic appraisals for the non-small-cell carcinomas. Because of its tendency to early metastases, however, small cell carcinoma is a special case. According to a study conducted by the American Joint Committe for Cancer Staging, the prognosis in this tumor is independent on the clinical stages I, II, and III. At the time of diagnosis, 84% of patients are already in the advanced stage III (Mountain and Hermes 1979). The TNM system is therefore less suitable for this tumor form than for the non-small-cell bronchial carcinomas. For this reason, an attempt was made to carry out a clinical classification with other staging systems. The most widespread is the concept of a classification suggested by the Veterans Administration Lung Cancer Study Group (Zelen 1973), whereby two stages are recognized: "limited disease" (restriction of the tumor to one hemithorax, including ipsilateral positive supraclavicular lymph nodes) and "extensive disease" (all forms of metastases beyond these boundaries) (Hansen et al. 1978; Rowitt et al. 1968) (Table 4). This classification was initially developed for radiotherapy, but was then also applied to chemotherapy and the concept of combined chemotherapy/radiotherapy of small cell bronchial carcinoma valid today. At the time of diagnosis,

Recent Results in Cancer Research. Vol. 97
© Springer-Verlag Berlin · Heidelberg 1985

Table 1. Classification of performance status according to Karnosky scale

Main classification	Score	Fine classification
Able to carry on normal activity; no special care is needed	100	Normal; no complaints; no evidence of disease
	90	Able to carry on normal activity; minor signs or symptoms of disease
	80	Normal activity with effort; some signs or symptoms of disease
Unable to work; able to live at home and care for most personal needs; a varying amount of assistence is needed	70	Cares for self; unable to carry on normal activity or to do active work
	60	Requires occasional assistence but is able to care for most of his needs
	50	Requires considerable assistence and frequent medical care
Unable to care for self; requires equivalent of institutional or hospital care; disease may be progressing rapidly	40	Disabled; requires special care and assistance
	30	Severely disabled; hospitalization is indicated although death not imminent
	20	Very sick; hospitalization necessary; active support treatment is necessary
	10	Moribund, fatal processes progressing rapidly
	0	Dead

Table 2. Zubrod performance scale (PS) suggested by the American Joint Committee

0	Fully active, able to carry on all predisease activities without restriction (Karnovsky 90−100)
1	Restricted in physically strenuous activity but ambulatory and able to carry out work of a light or sedentary nature, e.g., light housework, effice work (Karnovsky 70−80)
2	Ambulatory and capable of all self-care but unable to carry out any work activities. Up and about more than 50% of waking hours (Karnovsky 50−60)
3	Capable of only limited self-care; confined to bed or chair 50% or more of waking hours (Krnovsky 30−40)
4	Completely disabled. Cannot carry on any self-care. Totally confined to bed or chair (Karnovsky 10−20)

35%−45% of patients can be assigned to the stage "limited disease" (Osterlind et al. 1983).

On the basis of the possibilities of diagnosis available at present, a study group worked out a suggestion for a staging program on the occasion of a workshop of the International Association for the Study of Lung Cancer in 1981 (Osterlind et al. 1983). This suggestion is shown in Table 5. The group recommended that the extent of staging measures should

Table 3. TNM pretreatment clinical classification

T	Primary tumour
Tis	Preinvasive carcinoma (carcinoma in situ)
T0	No evidence of primary tumor
T1	Tumor 3 cm or less in its greatest dimension, surrounded by lung or visceral pleura and with no evidence of invasion proximal to lobar bronchus on bronchoscopy
T2	Tumor more than 3 cm in its greatest dimension of tumor of any size which, with its associated atalectasis or obstructive pneumonitis, extends to the hilar region. At bronchoscopy the proximal extent of demonstrable tumor must be at least 2 cm distal to the carina. Any associated atalectasis or obstructive pneumonitis must involve less than an entire lung and there must be no pleural effusion
T3	Tumor of any size with direct extension to adjacent structures such as the chest wall, diaphragm, or mediastinum and its contents, or tumor at bronchoscopy less than 2 cm distal to the carina, or tumor associated with atalectasis or obstructive pneumonitis of an entire lung or pleural effusion
TX	Any tumor that can be assessed or tumor proven by the presence of malignant cells in bronchopulmonary secretions but not visualized by radiography or bronchoscopy
N	Regional lymph nodes
N0	No evidence of regional lymph node involvement
N1	Evidence of involvement of peribronchial and/or homolateral hilar lymph nodes, including direct extension of the primary tumor
N2	Evidence of involvement of mediastinal lymph nodes
NX	The minimum requirements for assessment of the regional lymph nodes cannot be met
M	Distant metastases
M0	No evidence of distant metastases
M1	Evidence of distant metastases
MX	The minimum requirements for assessment of the presence of distant metastases cannot be met

The category M1 may be subdivided according to the following notation:

Pulmonary:	PUL	Bone marrow:	MAR
Osseous:	OSS	Pleura:	PLE
Hepatic:	HEP	Skin:	SKI
Brain:	BRA	Eye:	EYE
Lymph nodes:	LYM	Other:	OTH

depend on whether (a) surgical resection is foreseen; (b) the patient is to be included in a therapeutic study; or (c) an individual therapy is planned.

In the first two cases, the staging must be very extensive and all possible sites of metastasis must be considered. In the last case investigations can be limited to establishing the histological diagnosis, X-rays of the thorax, and specific investigations depending on the individual symptoms. In this situation, the subjective tolerance limit of the patient and the cost factor will have to be considered to a greater extent than in the other two situations. For therapeutic studies a precisely defined staging program is absolutely essential; without this it would not be possible to compare the results.

Table 4. Stages of spreading of small cell bronchial carcinomas

Limited disease	1. Primary tumor 2. Ipsilateral hilar lymph nodes 3. Ipsilateral supraclavicular lymph nodes 4. Ipsilateral and contralateral mediastinal lymph nodes 5. Possible presence of atelectasia 6. Paresis of the recurrent and/or phrenic nerve 7. Small-angle pleural effusion without malignant cells
Extensive disease	1. Contralateral hilar lymph nodes 2. Contralateral supraclavicular lymph nodes 3. Thorax wall infiltrations (also ipsilateral) 4. Pleuritis carcinomatosa, pleural effusion (apart from small-angle effusion without malignant cells) 5. Lymphangiosis carcinomatosa 6. Superior vena cava (SCV) syndrome 7. Metastases in the contralateral lungs 8. Other distant metastases (liver, brain, bone, other lymph nodes, etc.)

Limited Disease Sites

Diagnosis and Staging of the Primary Tumor

For evaluation of the primary tumor (T) and the regional lymph nodes, subdivision into a standardized basic diagnosis and some supplementary diagnostic techniques has proved to be appropriate (Table 6). They are supplemented by the diagnostic measures which clarify the operability when surgical resection may be planned and can exclude distant metastases.

The history is dominated by the main symptoms of bronchial carcinoma, such as dry cough, fever (retention pneumonia), night sweats, and hemoptysis, all of which also occur in other pulmonary diseases, e.g., tuberculosis. Further pointers to a bronchial carcinoma may be loss of weight, an abrupt fall in physical performance, chest pain, dyspnea, and paraneoplastic syndrome. Pain symptoms suggest that the T3 stage has already been reached with infiltration of the thorax wall (Table 3). As in the clinical symptoms, the bronchial carcinoma can also imitate any other lung disease in the X-ray (Grunze 1962). The scout-view X-rays taken in two planes constitute an essential part of the primary diagnostic procedure. They are supplemented by hilus tomograms. A paradoxical mobility of the diaphragm can be detected with fluoroscopy as a pointer to a central bronchial carcinoma with involvement of the phrenic nerve (T3 and/or N2).

Bronchoscopy has a central place among the diagnostic measures. Not only does it enable verification of the diagnosis in 60%−70% of patients, but it provides additional pointers to the T stage (e.g., distance of the tumor from the bifurcation) and the N stage (e.g., impression of the central airways and the trachea by lymph node metastases). Fiberoptic bronchoscopy substantially extends the possibilities of endoscopic investigation of bronchial carcinoma. Several authors have documented that computer tomography of the thorax is the most sensitive noninvasive method for reliable evaluation of the T stage. It enables any spreading of the tumor to the central areas to be determined, especially spreading to the hilus of the lung (Naidich et al. 1981a, b; Webb et al. 1981) and to the

Table 5. Staging procedures: Recommendations and comments (Osterlind et al. 1983)

Site	Procedure	Recommended	In special situations	Experimental	Comments
Primary tumor	Chest X-ray	+			
	Tomography		+		
	CT scan		+	+	Lacking specificity
	Fiberoptic bronchoscopy	+			
Mediastinum	Chest X-ray	+			
	Tomography		+		
	Mediastinoscopy		+		
	Gallium scintigraphy		+		
Pleura	Chest X-ray	+			
	Cytology	+			Of effusions
	Thoracoscopy		+		
Supraclavicular glands	Fine-needle aspiration	+			At least cytologic confirmation is recommended
	or biopsy		+		
Bone marrow	Marrow aspiration and biopsy	+			
	Bilateral		+		At least cytologic confirmation is recommended
	Scintigrams		+		
Liver	Peritoneoscopy and biopsy		+		
	Ultrasonography and biopsy		+	+	
	CT scans		+	+	At least cytologic confirmation is recommended
	Scintigrams		+		
CNS	CT scans		+		
	Scintigrams		+		
	Lumbal puncture		+		Positive cytology
	Myelograms		+		
Skin and lymph nodes	Fine-needle aspiration	+			At least cytologic confirmation is recommended
Retroperitoneal organs	Scintigrams			+	Too time-consuming
	CT scans		+	+	
	Ultrasonography		+	+	At least cytologic confirmation is recommended

Table 6. Diagnosis and staging of the primary tumor and regional lymph nodes

Basic diagnostic procedure
  Case history
  Clinical investigation and physical findings
  Laboratory investigations
  X-rays of the thoracic organs in two planes (fluoroscopy and tomography after recording findings)
  Bronchoscopy (bronchus lavage, catheter biopsy)

Supplementary diagnostic procedure
  Perfusion scintigraphy of the lungs
  Computer tomography
  Mediastinoscopy and needle biopsy
  Thoracoscopy
  Diagnostic thoracotomy

atrium of the heart. It also enables detection of the "blind regions" such as the costodiaphragmatic sinus, the retrocardial space, and the pleura (Huvenne et al. 1979; Margolis et al. 1974; Müller et al. 1981; Baron et al. 1982; Ekholm et al. 1982; Lackner et al. 1974; Schnyder and Gamsu 1981; Sommer et al. 1981; Wouters et al. 1982). Besides the precise size and position of the tumor, in particular an invasive growth into the thorax wall, excellent visualization of the mediastinum or the vessels of the hilus is possible with this method (Müller et al. 1981a, b) (Fig. 1). The bronchial system containing air is also accessible to evaluation (Naidich et al. 1981a, b; Webb et al. 1981).

The primary tumor and any secondary poststenotic alterations that may be present can also be visualized on conventional X-rays. For this reason, and in agreement with other authors (Harper et al. 1981; Ihde and Hansen 1981; Lewis et al. 1982), we consider that computer tomography is indicated for evaluation in the T stage in small cell bronchial carcinoma only for the patients in whom surgical resection is discussed or appears to be at all possible. There is a further indication for computer tomography of the thorax in specific cases in which tumor shadowing cannot be delimited with certainty from an inflammatory process by conventional radiology and bronchoscopy.

Scintigraphic investigations of the lungs with ^{99m}technetium, ^{81m}krypton, and ^{85m}xenon allow the functional parameters of perfusion and ventilation to be evaluated. With more advanced tumor stage or increasing tumor size, perfusion is reduced. However, this does not give any pointers with regard to operability or to extent of a planned operation (Ramos et al. 1974). However, scintigraphy permits precise preoperative determination of the ventilation reserves that will be left after the operation. For evaluation of the T stage, on the other hand, scintigraphy only gives indirect indications and is markedly inferior to other imaging techniques. According to Ramos et al. (1974), it allows correct determination of the T and N stages in 42% and 68% of cases, as against 68% and 80%, respectively, for conventional X-rays. Pulmonary perfusion scintigraphy continues to be indicated when there is any suspicion of a central bronchial carcinoma on the basis of the clinical radiology but this cannot be excluded or demonstrated by either bronchoscopy or radiology. Whereas 11% false-negative findings are recorded with X-rays of central bronchial carcinomas, a normal perfusion scintigram largely excludes a central bronchial carcinoma (only 2% false-positive findings) (Ramos et al. 1974).

The significance of pulmonary function scintigraphy (perfusion and ventilation) in small cell bronchial carcinoma, which is centrally located in the majority of cases, lies mainly in

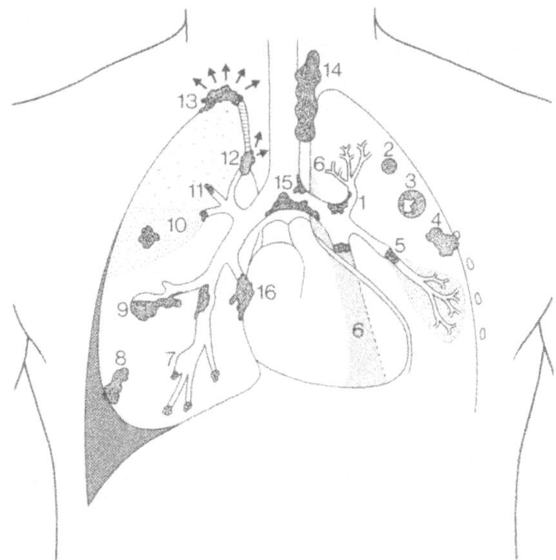

Fig. 1. Most frequent manifestation forms of a bronchial carcinoma (Grunze 1962): *1,* hilar lung cancer with endobronchial growth (relatively early elicitation of the cough reflex); *2,* typical round focus; *3,* tumor cavern (note the thick irregular walls); *4,* subpleural focus infiltrating the chest wall; *5,* obstructive segmental discontinuation with retention in pneumonia, in *10* already with abscess formation; *6,* atelectasis (hidden behind the heart shadow: lateral X-ray required); *7,* secondary bronchiectasia due to partial stenosis; *8,* focus near to the pleura with effusion; *9,* necrotizing tumor with draining bronchus (absecss symptom); *11,* obstruction emphysema due to valve occlusion; *12* and *13,* outbreak into the mediastinum, e.g., in the direction of the vena cava (upper inflow congestion) or as Pancoast tumor; *14,* lymph node involvement in the upper mediastinum and paratracheally, continuing up to the upper clavicular fossa (detection by lymph node biopsy according to Daniels or mediastinoscopy); *15,* carcinoma spreading to the trachea; *16,* carcinoma spreading to the pericardium. Caution: A bronchial carcinoma can be masked even in a normal X-ray

that it enables the result of therapy to be determined on the basis of functional parameters. Alterations in tumor size can be demonstrated better by conventional radiography, but are not necessarily accompanied by an improvement or deterioration of ventilation or perfusion.

The ideal combination of functional and morphological diagnostic techniques is doubtless the joint application of computer tomography and lung function scintigraphy (Herter and König 1983). The result of each of these investigations should be interpreted with reference to the other. This combination allows important statements with regard to tumor position and operability in certain circumstances, which are particularly relevant when a surgical operation is planned.

When conventional radiodiagnostics or a computer tomogram have provided indications of tumor infiltration or metastatic spread into the pleura an aspiration of the pleural fluid must be carried out. Detection of tumor cells would mean that the stage of extensive disease has already been reached. A small pleural effusion without demonstration of pathologic cells, e.g., resulting from lymphatic congestion in a central tumor, would still justify classification as limited disease. However, as the sole symptom of dissemination, the pleural effusion has only a slight prognostic significance (Linvingston et al. 1981) In a pleural effusion of unknown genesis, thoracoscopy as a supplementary diagnostic measure occasionally results in diagnosis of a small cell bronchial carcinoma. However, this method of investigation is not used as a routine measure in staging of this tumor.

Regional Lymph Node Metastases

When there is involvement of the regional lymph nodes, the chances of cure in a patient with a non-small-cell bronchial carcinoma are drastically, reduced (Martini 1979; Mountain and Hermes 1979). This deterioration in prognosis is not so pronounced in small-cell bronchial carcinoma, because the prognosis is poorer in general and because there is early dissemination via the blood. However, it is very important to know about any involvement of the hilar and mediastinal lymph nodes in small cell bronchial carcinoma when surgical resection is under discussion in the early stage. This would no longer be feasible when metastasis has occurred beyond the first lymph node level. Involvement of the contrahilar lymph node region would mean that the stage of extensive disease has been reached.

It has become possible to evaluate hilar lymph nodes by X-ray and hilus tomograms (Müller et al. 1981; Osborne et al. 1982), and possibly also by bronchoscopy. The extensive involvement of the mediastinal lymph nodes can be assumed in radiologically demonstrated widening of the mediastinum. Computer tomography is this region (Fig. 2) allows earlier detection of metastatic involvement of the lymph nodes than conventional X-ray diagnostics (Bähren et al. 1982; König et al. 1983; Vock and Haertel 1981). Histological verification of involvement of this region is possible by mediastinoscopy. It has a firm place in staging of non-small-cell bronchial carcinoma (Lüllig et al. 1977; Martini 1979). In small cell bronchial carcinoma, this investigation only has a role in staging when surgical resection is planned. This is the exception. For discrimination between the stages of limited and extensive disease mediastinoscopy is not required (Lunia et al. 1979). This lymph node region is affected in most patients (Zehlen 1973).

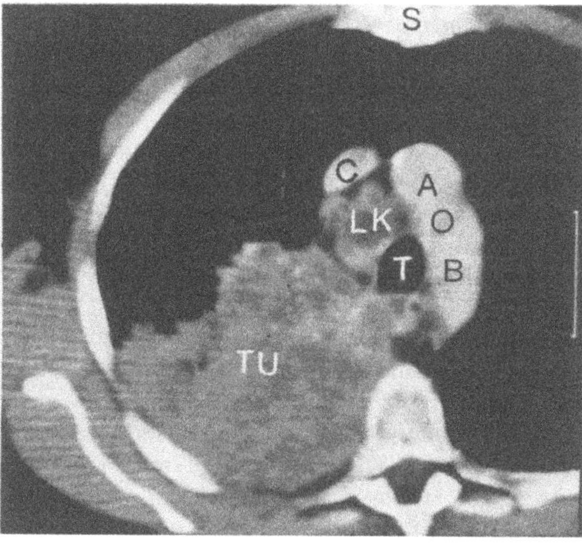

Fig. 2. Large bronchial carcinoma (*TU*) of the right upper lobe with metastases of the mediastinal lymph nodes *(LK)*. The tumor extends from the dorsolateral thoracic wall to the esophagus *(O)*. (*S*, sternum; *C*, vena cava superior; *T*, trachea; A O B, aortic arch)

According to Maurer et al. (1980) and Cooksey et al. (1979), metastatic spread into the supraclavicular regions has no great prognostic significance. There is hence a school of thought that even patients with contralateral supraclavicular lymph node metastases should be assigned to the stage of limited disease (Osterlind et al. 1983). When an enlargement of the supraclavicular lymph nodes can be palpated this finding must be verified by fine-needle aspiration or biopsy to check for metastasis.

Extensive Disease Sites

Small cell bronchial carcinoma has frequently already undergone distant metastatic spread at the time of diagnosis. For this reason, screening for distant metastases in part of the primary diagnosis program. It is concentrated on the most frequent sites of distant metastases: liver, skeleton, brain, and adrenals. In small cell carcinoma the abdominal lymph nodes (57%), the pancreas (31%), and the kidneys (22%) are also involved in metastatic spread in a high percentage of cases (Hansen et al. 1978). These figures are based on the results of autopsies (Selawry and Hansen 1982). However, they can also be largely confirmed by an extensive program of clinical investigations even at the time of primary diagnosis.

Heaptic Metastases

The liver is frequently affected by metastases of a bronchial carcinoma. This applies particularly in the case of small cell carcinoma (Figs. 3–5). Besides the results of clinical investigation, the results of clinical laboratory investigations (increase of the activities of SGOT, GGT, AP, and LDH) indicate involvement of the liver. These laboratory parameters can of course also be pathologically altered by concomitant diseases. They should therefore only be interpreted as a pointer. Dombernowsky et al. (1978) recorded a

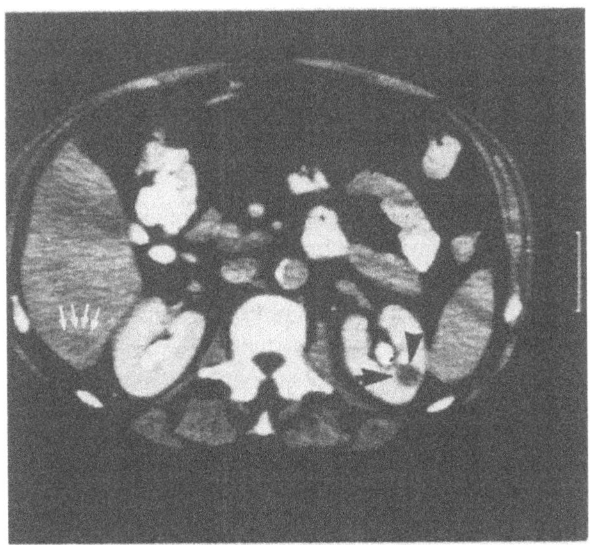

Fig. 3. Metastases of a small cell lung cancer in the left kidney *(black arrows)* and the liver *(white arrows)*

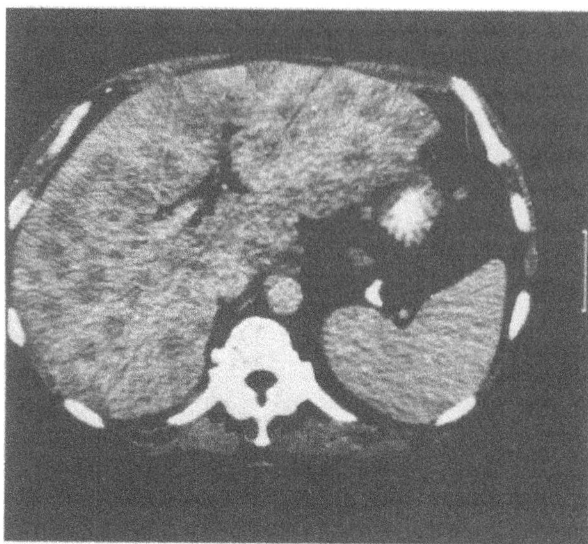

Fig. 4. Multiple small spotted liver metastases in a small cell lung cancer

Fig. 5. Sonographic evidence of multiple sonographically silent liver metastases *(white arrows)* in small cell lung cancer

very much higher percentage of pathologic liver parameters in an investigation series of 190 patients with small cell bronchial carcinoma in 34 patients with histologically verified liver metastases than in patients without liver metastases (alkaline phosphatase 71% as opposed to 14%, SGOT 56% as opposed to 0%, LDH 79% as opposed to 16%).
The noninvasive imaging methods available for investigation of the liver include scintigraphy, sonography, and computer tomography. These techniques complement each other. Each provides information that cannot be expected from the other. Their application depends on the personal experience of the investigator and the technical prerequisites. The results obtained with these techniques require cytological or histological confirmation to be interpreted with certainty as tumor specific (Lee 1978; Ferrucci 1979; Bruyan et al. 1977). This is achieved by fine-needle aspiration or biopsy.

Because it s simply and rapidly carried out, scintigraphy has been regarded by many clinicians in recent years as the preferred method of investigation. It gives an overview of the tissue capable of functioning, but as a routine screening method it does not proved any guaranteed appraisal because of false-positive results and the incapacity to detect foci with a diameter of less than 2 cm (Hansen et al. 1978; Muggia and Chervu 1974; Petasnick et al. 1979). When a doubtful or definite pathologic finding is obtained by scintigraphy, it can be frequently clarified by means of sonography or computer tomography. These two investigations are especially helpful for evaluation of the left lobe of the liver and the hilum (Petasnick et al. 1979).

Sonography, with its appreciable technical developments in recent years, is equal as a screening method to scintigraphy for the detection of circumscribed foci in the liver, and is indeed superior for evaluation of the left lobe of the liver. However, it requires an experienced and meticulous investigator and a long duration of the investigation. Compared with other investigation techniques, the advantages of sonography consist in the absence of stress and risk for the patient, the frequent repeatability, and the possibility of preparing longitudinal profiles and of looking at dubious areas in three dimensions. It permits definite distinction between cysts and solid foci. However, sonography is limited by pronounced adiposity and by flatulence in the colon and surrounding bony structures. Hence, parts of the liver which are situated directly under the costal arch or covered by the lungs or the colon can be evaluated only with difficulty. Schwerk and Schmitz-Moormann (1981) succeeded in attaining a diagnostic specificity of 90% by needle biopsies guided ultrasonically in 60 patients with malignant tumors and liver findings demonstrated by ultrasonography.

Snow et al. (1979) and Wallace and Chuang (1982) compared the results of scintigraphy, sonography, and computer tomography for the evaluation of liver metastases. The authors calculated the sensitivity and specificity as 96% and 86% for computer tomography, and as 94% and 67%, respectively, for scintigraphy. The comparatively poorer results of sonography (75% and 50%, respectively), which were published in 1979, should be appreciably improved by the use of modern ultrasonographs such as are available today.

In the course of recent years, due to the development of instruments which allow short durations of exposure, computer tomography has gained importance in screening for liver metastases. The sole disadvantage is the enormous degree of technical sophistication required, which is associated with high costs. Pathologic computer tomograms of the upper abdomen were recorded in 10%−37% of small cell bronchial carcinoma patients by Vas et al. (1981) and by Poon et al. (1982). In most cases the liver and the adrenals were affected by metastases. Even though computer tomography has appreciably facilitated diagnosis of tumors and metastases in the abdominal area, there is still no clarity at present with regard to the necessity of routine pretreatment investigation of the organs of the upper abdomen in small cell bronchial carcinoma by means of computer tomograms, apart from in scientific studies. Although computer tomography can differentiate and demonstrate space-occupying processes (absecesses, benign and malignant tumors, hemangiomas, adenomas, infarctions) with greater certainty than echography, routine staging of the upper abdomen (liver, kidneys, spleen, adrenals, and retroperitoneum) can doubtless be carried out more cheaply by means of sonography. Ihde et al. (1980) carried out computer tomography of the liver with subsequent peritoneoscopy in 66 patients. They attained a specificity of 92% and a sensitivity of 64%.

When computer tomography is not able to answer beyond all doubt the question as to metastatic spread of bronchial carcinoma into the liver, peritoneoscopy is indicated. It had already been used successfully by several study groups (Dombernovsky et al. 1978; Muggia and Chervu 1974; Margolis et al. 1974) before the era of computer tomography to look for metastases in small cell bronchial carcinoma. Peritoneoscopy is able to reveal smaller foci, when they are situated on the liver surface and in the area screened, than those that can be detected with the other imaging techniques described. The investigation has the advantage of histological confirmation of the diagnosis and appraisal of contiguous structures.

Skeletal and Bone Marrow Metastases

The skeleton is a further predilection site for metastatic spread of small cell bronchial carcinoma. The involvement is very often demonstrated even clinically (Muggia and Chervu 1974) by bone marrow aspiration or biopsy, scintigraphy (Kies et al. 1978; Wilson and Calhoun 1981; Pabst and Langhammer 1977) and X-ray before skeletal metastases are indicated by pain, a pathologic blood count, and elevation of the serum calcium or alkaline phosphate. Metastatic spread into the bone marrow signifies a poor prognosis (Hirsch and Hansen 1980).

Nonspecifically guided needle biosy at the dorsal cranial iliac crest yields positive results in the phase of primary diagnostics in up to 30% of patients with small cell bronchial carcinoma (Hansen et al. 1978; Hirsch et al. 1977; Präuer et al. 1975). We were able to demonstrate metastatic spread into the bone marrow by means of this method in only 5% (3/57) in primary staging of untreated patients (Krempien et al. 1980). Hirsch et al. (1979) demonstrated that the precision of this method can be increased in bilateral biopsy of the dorsal iliac crests. Bone marrow aspiration and biopsy have a greater precision in small cell bronchial carcinoma than scintigraphy and X-ray (Hansen 1974). Hirsch et al. (1977) consider that, in contrast to other tumors, the cytological investigation of the aspirated bone marrow is superior to histological evaluation of a bone marrow cylinder in small cell bronchial carcinoma. Cytological and histological investigation of the bone marrow must therefore always be contained in the primary staging program of small cell bronchial carcinoma.

The bone scintigram with ^{99m}technetium is used as a method of screening for skeletal metastases in asymptomatic patients. The method has an exceedingly high sensitivity (Pistenma et al. 1975), which is not reached by any other noninvasive technique in screening for skeletal metastases. However, it displays a very low specificity. An accumulation of the radioactive tracer must be regarded as a nonspecific finding (e.g., abscess, primary bone tumor, metastasis, degenerative alterations). For this reason, a suggestive finding therefore requires a subsequent radiological check. In this way, abscesses, primary bone tumors, and degenerative alterations can be excluded in most cases. However, a negative X-ray does not necessarily mean the absence of metastases, since these can be only demonstrated by conventional radiology when at least 50%−75% decalcification is demonstrable or when the focus has attained a diameter of 1−1.5 cm (Edelstyn et al. 1967). The interpretation of a dubious skeletal finding in the scintigram is frequently only possible retrospectively, since scintigraphic alterations can precede the radiologically demonstrable destruction and only the latter is definitive.

CNS Metastases

Metastases of the CNS are of increasing clinical and prognostic significance (Jacobs et al. 1977). Their incidence is reported as 20%−50% after autopsy investigations and

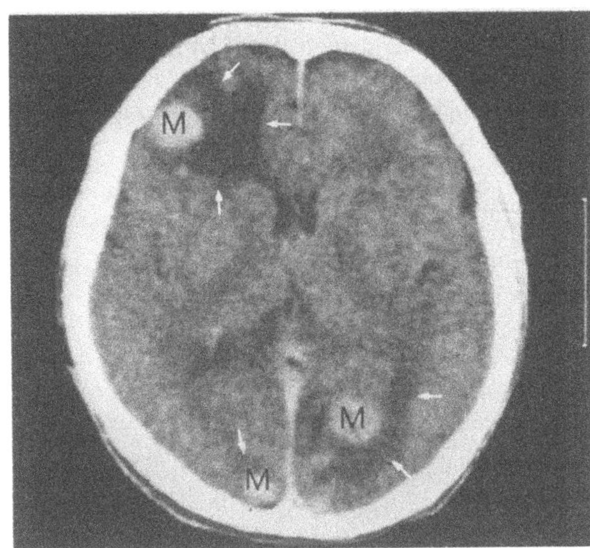

Fig. 6. Multiple brain metastases *(M)* of small cell lung cancer, with extensive collateral edema *(white arrows)*. One of the metastases is situated in the left frontal brain, and two others are in the right and left occipital lobes

14%–30% after clinical investigations (Bunn et al. 1978; Selawry and Hansen 1982). With the increased survival of patients due to improved therapy, metastasis to the CNS has increased appreciably in clinical significance in small cell bronchial carcinoma. It is taken into account in the therapy program in the form of prophylactic irradiation of the CNS.

CNS metastases are observed especially in the brain (Fig. 6) but also in the hypophysis, the leptomeninges, and the spinal cord (Bunn et al. 1978). They are diagnosed clinically in 80% of cases. The remaining 20% are only discovered at autopsy. In contrast to primary brain tumors, an abrupt onset of symptoms is characteristic for the brain metastases. The most frequent symptoms are headaches, alterations of mental behavior, and a circumscribed motor weakness. In addition, vertigo, ataxia, and aphasia occur. The symptoms of involvement of the meninges are similar. Cranial nerve symptoms are then frequently observed in addition. Metastatic spread into the spinal cord is manifested initially by pain, and later by motor weakness, distrubance of sensitivity, and dysfunctions of the autonomic nervous system. The picture of a transverse lesion syndrome can develop within days to weeks.

In the symptomatic stage, besides clinical and neurological examination, computer tomography, the electroencephalogram, lumbal puncture with cytological investigation of the CSF, and myelography are available for diagnosis. Depending on the manifestation of the metastasis, these methods vary in different precision. Brain scintigraphy now has only historical importance as a method of screening for brain metastases. It has been replaced by computer tomography (Poon et al. 1982). No other technique has the same specificity or sensitivity as a method of screening. With computer tomography, brain metastases are demonstrated in 10%–14% of the patients in routine investigations (Burges et al. 1979; Nugent et al. 1979). This incidence could be confirmed in a retrospective analysis we carried out ourselves in 350 patients with small cell bronchial carcinoma. In 11 patients, brain metastases were demonstrated by the computer tomogram during the primary staging, and in 37 further patients in the later course. According to Johnson et al. (1983), one should distinguish between symptomatic and asymptomatic patients. The authors demonstrated brain metastases in the computer tomogram in 81% of the patients with corresponding symptoms. In asymptomatic patients a frequency of metastatic spread of only 1.6% could be demonstrated by means of routine computer tomography of the brain. It may be concluded from this that a computer tomogram of the brain need not be demanded before the institution of therapy outside scientific clinical studies.

Metastases in Other Regions

Among the remaining regions, the retroperitoneal space has great significance for metastatic spread of small cell bronchial carcinomas. According to autopsy findings,

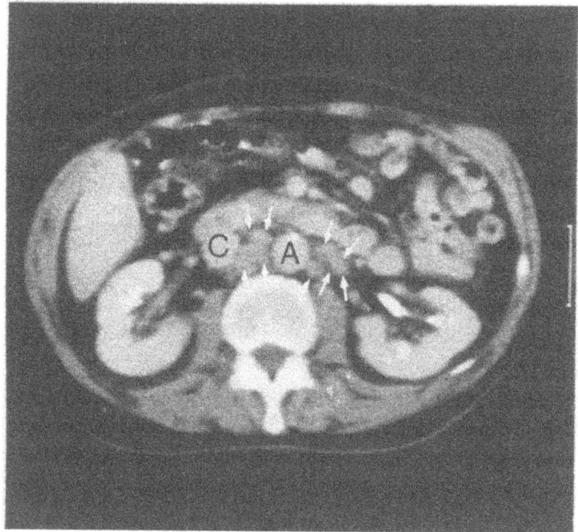

Fig. 7. Invasion of the retroperitoneal lymph nodes *(white arrows)* by metastases of a small cell lung cancer. The vena cava superior *(C)* and the aorta abdominalis *(A)* are distended by enlarged lymph nodes *(white arrows)*

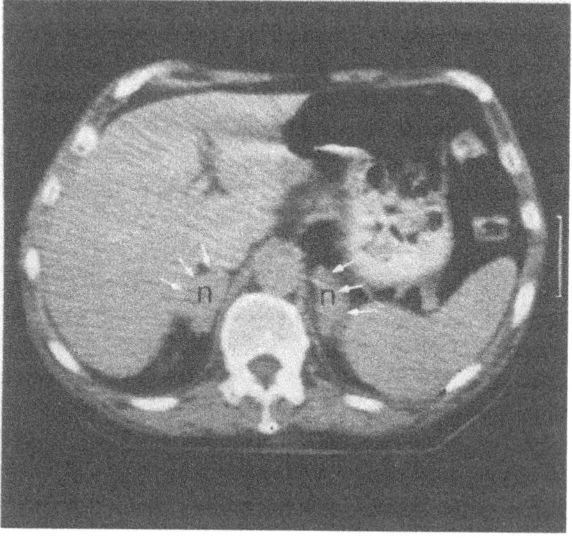

Fig. 8. Metastases of a small cell lung cancer in both adrenals *(n)*, which are markedly enlarged *(white arrows)*

metastases can be demonstrated in the retroperitoneal lymph nodes in 57% of patients (Fig. 7) and in the adrenals in 39% of patients (Fig. 8) (Hansen 1974). Evaluation of this region is possible with sonography, but even better with computer tomography without great stress to the patient (Jaschke et al. 1980; Poon et al. 1982; Vas et al. 1981; Ihde et al. 1980). Because of the high incidence of metastasis in the retroperitoneal space, one of these two investigation techniqeus should be included in the routine program for staging of small cell bronchial carcinoma.

Differential diagnosis between metastases and adenomas can give rise to difficulties (Sandler et al. 1982), since a distinction between metastases and adenomas is not possible in space-occupying processes between 1 and 3 cm in diameter. In larger, central necrotic processes metastases are involved in most cases. The hormone parameters are likewise unable to give any further indications in inactive adenomas or in metastases. In addition, it must be taken into account that precisely in small cell bronchial carcinoma paraneoplastic ACTH production can frequently be present, with secondary hypertrophy of the adrenals. This can be demonstrated very well by computer tomography and can be delimited from other adrenal processes. According to investigations of Pedersen et al. (1982), on the other hand, computer tomography does not allow any distinction between metastasis or hypertrophy in patients with small cell bronchial carcinomas and unilateral or bilateral adrenal enlargement. The authors do not specify their computer tomographic criteria. A differentiation of hypertrophy, unilateral adenomas, and metastases is possible as a rule on the basis of hormonal investigations.

Restaging

Repeated investigations are absolutely necessary evaluate the disease course and the result of therapy. This is defnied according to the criteria of the World Health Organization (1979). In this way it is possible to compare the results.

The intensity and frequency of the investigations in restaging influence the quality of the appraisal with regard to the degree and the duration of remission. For this reason, a detailed program must be laid down for scientific clinical studies. Outside clinical studies one will be able to restrict investigations to a minimum program comprising repeated investigation of all initially pathologic parameters. X-Rays of the thorax before each therapy cycle are absolute necessary. In a prospective therapeutic study (Harms et al. 1983), we have already confirmed the prognostic significance of this investigation technique after the first therapy cycle. Patients in whom there was no reduction of the tumor mass had a significantly shorter survival than the patients who had responded to the first therapy cycle. It may be concluded from this that the therapeutic procedure has to be changed after the first treatment cycle when the tumor mass shows stationary behavior.

For definition of a complete remission in the area of the primary tumor a fiberoptic bronchoscopy with histological investigation is indispensable. Mediastinoscopy will be necessary when additional local therapy (e.g., radiotherapy) is to be dispensed with after attainment of complete remission.

The investigation program for restaging is based on the pretreatment staging. A corresponding suggestion was elaborated by a study group (Osterlind et al. 1983) of the International Association for the Study of Lung Cancer (Table 7). We endorse this suggestion.

Table 7. Recommendations for restaging (Osterlind et al. 1983)

Site	Procedure	Recom-mended	If positive initially	If signs/symptoms
Primary	Chest X-ray	+		
	Fiberoptic bronchoscopy	+		
Mediastinum	Mediastinoscopy[a]	+		
	Gallium scan		+	
Bone marrow	Biopsy and aspiration	+		
	Bilateral biopsies		+	
	Scintigrams		+	
Liver	Peritoneoscopy and liver biopsy		+	
	Ultrasonography		+	
	CT scan		+	
Lymph nodes and skin	Fine-needle aspiration			+
CNS	CT scans		+	
	Scintigrams		+	
	Lumbal puncture		+	
	Myelograms		+	
Retroperitoneal organs	CT scans		+	+
	Ultrasonography	+	+	
	Laparotomy			+

[a] Whenever possible

Persistent lesions should if possible be evaluated histologically or cytologically. It is conceivable that non-small-cell portions of the tumor might be detected, which require another therapy.

In the course of therapy, investigations are restricted to the interim case history, physical examination, clinical laboratory and blood chemistry investigations, and scout-view X-rays of the thoracic organs. Further investigations are conducted in accordance with the individual extent of the tumor or the symptoms of the patient. Before the end of therapy an extensive staging will be carried out, taking account of all initial phatologic findings and current symptoms.

In the case of complete remission the patient must be examined every 4 weeks in the first year, every 3 months for a further 2 years, and finally every 6 months in the context of the follow-up. These investigations comprise the interim case history, physical examination, blood chemistry values, a scout-view X-ray of the thorax in two planes, and further investigations depending on individual symptoms. Regular and meticulous examination of the patient is especially important in that small cell bronchial carcinoma tends to develop recurrences rapidly. In addition, the long-term disease-free survivors can only be detected in this way.

Small cell bronchial carcinoma tends to be accompanied by ectopic synthesis of peptide hormones or biomarkers, which can induce paraneoplastic syndromes (Gropp et al., this volume). When these hormones or biological markers are determined before treatment their analysis should be continued regularly in the context of scientific clinical studies. In this way, it might be possible to establish whether they are of prognostic importance, like other tumor manifestations, and whether they allow earlier diagnosis of a recurrence. Preliminary investigations so far (Drings et al. 1983) reveal a behavior which is largely consistent with other manifestations.

References

American Joint Committe for Cancer Staging and Reporting (1979) Staging of lung task force on lung cancer. Chicago

Bähren W, Sigmund G, Lenz M (1982) Wertigkeit der Computertomographie im Vergleich zu Mediastino- und Probethorakotomie bei intrathorakalen Tumoren mit mediastinaler Beteiligung. Fortschr Röntgenstr 137: 269−274

Baron RL, Levitt RG, Sagel SS, White MJ, Roper CL, Marbarger JP (1982) Computed tomography in the preoperative evaluation of bronchogenic carcinoma. Radiology 145: 727−732

Bruyan PJ, Dinn WM, Grossmann ZD (1977) Correlation and computed tomography, gray scale ultrasonography and radionucleide imaging of the liver in detecting spaceoccupying processes. Radiology 124: 387−393

Bunn PA, Nugent J, Matthews MJ (1978) Central nervous system metastases in small cell bronchogenic carcinoma. Semin Oncol 5: 314−322

Burgess RE, Burgess VF, Dibella NJ (1979) Brain metastases in small cell carcinoma of the lung. J Am Med Assoc 242: 2084−2086

Carr DT (1973) Diagnosis, staging and criteria of response to therapy for lung cancer. Cancer Chemother Rep Pat 3: 303−305

Carter SK (1979) Introduction − What has happend in the last five years. In: Muggia F, Rozencweig M (eds) Lung cancer: Progress in therapeutic research. Raven, New York, pp 1−12

Cooksey JA, Bitran JD, Desser RK et al. (1979) Small-cell carcinoma of the lung: the prognostic significance of stage on survival. Eur J Cancer 15: 859−865

Dombernowsky P, Hirsch F, Hansen HH, Hainau B (1978) Peritoneoscopy in the staging of 190 patients with small-cell anaplastic carcinoma of the lung with special reference to subtyping. Cancer 41: 2008−2012

Drings P, Manke HG, Holle R, Victor N, Havemann K, Harms V, Gropp C, Thomas C, Kalbfleisch H, Bischof W, Hans K, Westerhausen M, Schroeder M, Queißer W, Heim M, Dirks P, Häßler R, Becker H, Gayer G, Wehr E, Meuret G, Mende S, Pralle H, Graubner M, Löffler H, Pfannschmidt G, Georgii A, Diehl V, Weiß J, Mitrou PS, Klippstein TH (1983) The tumor markers CEA, calcitonin and ACTH in a prospective trial of small cell lung cancer (SCLC). Proceedings of the 13th international congress of chemotherapy. Vienna, pp 274/45−274/48

Edelstyn GA, Gillespie PJ, Grebell FS (1967) The radiological demonstration of osseous metastasis: Experimental observations. Clin Radio 18: 158

Edmonson JH, Lagakos SW, Selawry OS, Perlia CP, Bennett JM, Muggia FM, Wampler G, Brodovsky HS, Horton J, Colsky J, Mansour EG, Creech R, Stolbach L, Greenspan EM, Levitt M, Israel L, Ezdinli E, Carbone P (1976) Cyclophosphamide plus CCNU compared to cyclophosphamide alone in the treatment of small cell and adenocarcinoma of the lung. Cancer Treat Rep 60: 925−932

Ekholm S, Albrechtson U, Kugelberg J, Thylen U (1982) Computed tomography in preoperative staging of bronchogenic carcinoma. J Comput Assist Tomogr 6: 763−765

Ferrucci JT (1979) Body ultrasonography. II. N Engl J Med 300: 590−602

Grunze H (1962) Tumoren der Thoraxorgane. In: Bartelheimer H, Maurer HJ (eds) Diagnostik der Geschwulstkrankheiten. Thieme, Stuttgart

Hansen HH (1974) Bone metastases in lung cancer. Munksgaard, Copenhagen

Hansen HH, Dombernowsky P (1981) Small cell anaplastic carcinoma of the lung: staging. In: Livingston RG (ed) Lung cancer 1. Martims Nijhoff, The Hague, pp 157−168

Hansen HH, Dombernowsky P, Hirsch FR (1978) Staging procedures and prognostic features in small-cell anaplastic bronchogenic carcinoma. Semin Oncol 5: 280−287

Harms V, Havemann K, Drings P, Holle R, Gropp C, Victor V, Manke HG, Hnas K, Westerhausen M, Schroeder M, Thomas C, Kalbfleisch H, Bischof W, Queißer W, Heim M, Dirks P, Häßler R, Becker H, Wehr E, Bayer G, Meuret G, Mende S, Pralle H, Graubner M, Löffler H, Pfannschmidt G, Georgii A, Diehl V, Weiß J, Mitrou PS, Klippstein T (1983) A randomized multicenter trial comparing sequential versus alternating polychemotherapy in small cell lung cancer. Proceedings of the 13th international congress of chemotherapy, Vienna, pp 248/10−248/14

Harper PG, Houang M, Spiro SS, Geddes D, Hodson M, Souhami RL (1981) Computerized axial tomography in the pretreatment assessment of small-cell carcinoma of the bronchus. Cancer 47: 1775−1780

Herter M, König R (1983) Computertomographie und Lungenfunktionsszintigraphie beim zentralen Bronchialcarcinom: Vergleich zwischen morphologischen und funktionellen Veränderungen. Fortschr Röntgenstr 138: 707−715

Hirsch FR, Hansen HH, Dombernowsky P, Hainau B (1977) Bone marrow examination in the staging of small-cell anaplastic carcinoma of the lung with special reference to subtyping. An evaluation of 203 consecutive patients. Cancer 39: 2563−2567

Hirsch FR, Hansen HH, Hainau B (1979) Bilateral bone-marrow examinations in small cell anaplastic carcinoma of the lung. Acta Pathol Microbiol Scand 87: 59−62

Hirsch FR, Hansen HH (1980) Bone marrow involvement in small-cell anaplastic carcinoma of the lung. Prognosis and therapeutic aspects. Cancer 46: 206−211

Huvenne R, Girvegnee A, Kuhn G, Bollaert A (1979) Performances of IIId generation scanner in the diagnosis and assessment of loco-regional extension of lung tumor. EORTC Symposium on Progress and Perspection in Lung Cancer Treatment, Brussels

Ihde DC, Hansen HH (1981) Staging procedures and prognostic factors in small cell carcinoma of the lung. In: Greco FA, Oldham RK, Bunn PA (eds) Samall cell lung cancer. Grune and Stratton, New York, pp 261−283

Ihde DC, Dunnick NR, Johnstom-Early A et al. (1980) Abdominal computed tomography in small cell lung cancer: assessment of extent of disease and response to therapy. Clin Res 48: 416

Jacobs L, Kinkel WR, Vincent RG (1977) "Silent" brain metastases from lung carcinoma determined by computerized tomography. Arch Neurol 34: 690−693

Johnson DH, Windham WW, Allen JH, Greco FA (1983) Limited value of CT brain scans in the staging of small cell lung cancer. AJR 140: 37−40

Jaschke W, van Kaick G, Palmtag H (1980) Vergleich der Wertigkeit von Echographie und Computertomographie bei der Diagnostik raumfordernder Prozesse der Nieren. Fortschr Röntgenstr 132: 145−151

Karnofsky D, Abelmann WH, Craver LF et al. (1948) The use of nitrogen mustards in the palliative treatment of carcinoma with particular reference to bronchogenic carcinoma. Cancer 1: 634−656

Kies MS, Baker AW, Kennedy PS (1978) Radionuclide scans in staging of carcinoma of the lung. Surg Gynecol Obstet 147: 175−176

König R, van Kaick G, Lüllig H, Vogt-Moykopf I (1983) Computertomographische Beurteilung mediastinaler Lymphknoten beim Bronchialcarcinom. Fortschr Röntgenstr 138: 682−688

Krempien F, Dirks HP, Krempien B, Drings P (1980) Metastasierung in das Knochenmark und Osteopathie bei kleinzelligem Bronchialkarzinom. Blut 41: 302

Lackner K, Felix R, Oeser H, Wegener OH, Bücheler E, Heuser L, Mödder U, Thurn P (1974) Erweiterung der Röntgendiagnostik im Thoraxbereich durch die Computertomographie. Radiologie 19: 79−89

Lee YTN (1978) Liver scanning in patients with malignancy: rationale, results, and limitations. Cancer Treat Rep 62: 1183−1191

Lewis R, Bernardino ME, Valdivieso M, Farha P, Barnes PA, Thomas JL (1982) Computed tomography and routine chest radiography in oat cell carcinoma of the lung. J Comput Assist Tomogr 6: 739−745

Livingston RB, McCracken JD, Chen T (1981) Isolated pleural effusion in small lung cell carcinoma: favorable prognostic group. Proc Am Assoc Cancer Res ASCO 22: 200

Lüllig H, Hewera K, Vogt-Moykopf I (1977) Die Mediastinoskopie − Indikation und Aussagefähigkeit. Prax Pneumol 31: 25−28

Lunia SL, Buckdeschel J, McKneally MF et al. (1979) Mediastinal involvement by bronchogenic carcinoma: correlation readioterhapy, gallium scanning and surgical biopsy. Proc Am Assoc Cancer Res ASCO 20: 405

Margolis R, Hansen HH, Muggia FM, Kanhouwa S (1974) Diagnosis of liver metastases in bronchogenic carcinoma. A comparative study of liver scans, function tests, and peritoneoscopy with liver biopsy in 111 patients. Cancer 34: 1825–1928

Martini N (1979) Identification and prognostic implications of medistinal lymph node metastases of the lung. In: Muggia F, Rozencweig M (eds) Lung cancer: Progress in therapeutic research. Raven, New York, pp 251–255

Maurer LH, Tulloh M, Weiss RB et al. (1980) A randomized combined modality trial in small cell carcinoma of lung. Cancer 45: 30–39

Mountain CF, Hermes KE (1979) Management implications of surgical staging studies. In: Muggia F, Rozencweig M (eds) Lung cancer: Progress in therapeutic research. Raven, New York, pp 233–242

Muggia FM, Chervu LR (1974) Lung cancer: diagnosis in metastases sites. Semin Oncol 1: 217–228

Müller HA, van Kaick G, Schaaf J, Lüllig H, Vogt-Moykopf I (1981) Präoperatives Staging des Bronchialcarcinoms: Wertigkeit der Computertomographie im Vergleich zur konventionellen Radiologie. Fortschr Röntgenstr 134: 601–607

Naidich DP, Khouri NF, Scott WW Jr, Wand KP, Siegelmann SS (1981a) Computed tomography of the pulmonary hila. 2. Abnormal anatomy. J Comput Assist Tomogr 5: 459–467

Naidich DP, Khouri NR, Stitik FD, McCauley DI, Siegelmann SS (1981b) Computed tomography of the pulmonary hila. 2. Abnormal anatomy. J Comput Assist Tomogr 5: 468–475

Nugent JL, Bunn PA, Matthew MJ (1979) CNS metastases in small cell bronchogenic carcinoma. Cancer 44: 1885–1892

Osborne DR, Korobkin M, Ravin CE, Puman CE, Wolfe WG, Searly WC, Youg WG, Breitman R, Heaston D, Pam P, Halber M (1982) Comparison of plain radiology, conventional tomography and computed tomography in detecting intrathoracic lymph node metastases from lung carcinoma. Radiology 142: 157–161

Osterlind K, Ihde DC, Ettinger DS, Gralla RJ, Karrer K, Kraus S, Maurer LH, Rorth M, Sörenson S, Vincent R (1983) Staging and prognostic factors in small cell carcinoma of the lung. Cancer Treat Rep 67: 3–9

Pabst HW, Langhammer H (1977) Detection and differential diagnosis of bone lesions by scintigraphy. Eur J Nucl Med 2: 261–268

Pedersen AG, Hansen M, Kehlet H et al. (1982) Adrenocortical function compared with CT-scanning in small cell anaplastic carcinoma of the lung. Acta Radiol 20: 353–355

Petasnick JP, Ram P, Turner DA, Fordham EW (1979) The relationship of computed tomography, gray-scale ultrasonography and radionucleide imaging in the evaluation of hepatic masses. Semin Nucl Med 9: 8–21

Psitenma DA, McDougall IR, Kriss JP (1975) Screening for bone metastases. Are only scans necessary? J Am Med Assoc 231: 46–50

Poon PY, Feld R, Evans WK, Ege G, Yech JL, McLoughlin ML (1982) Computed tomography of brain, liver, and upper abdomen in the staging of small cell carcinoma of the lung. J Comput Assist Tomogr 6: 963–965

Präuer HW, Rastetter J, Sauer E, Ultsch B (1975) Ergebnisse der ungezielten Beckenkamm-Biopsie bei Patienten mit Bronchus-Karzinom. Münch Med Wochenschr 117: 1821–1824

Ramos M, Baumgartner MW, Rosler H (1974) Der Wert der 133 Xe-/99m Tc-MAP-Lungen-szintigraphie für Früherkennung und präoperative Abklärung des zentralen Bronchialcarcinoms. Ther Umsch 31: 731–737

Rowitt B, Patno ME, Rapp R (1968) The survival of patients with inoperable lung cancer: a large scale randomized study of radiation therapy versus placebo. Radiology 90: 688

Sandler MA, Pearlberg JL, Madrazo BL, Gitschlag KF, Gross SC (1982) Computed tomography evaluation of the adrenal gland in the preoperative assessment of bronchogenic carcinoma. Radiology 145: 733–736

Schnyder PA, Gamsu G (1981) CT of the pretracheal retrocaval space. Am J Roentgenol 136: 303–308

Schwerk WB, Schmitz-Moormann P (1981) Ultrasonically guided fine-needle biopsies in neoplastic liver disease. Cancer 48: 1469–1477

Seldwry OS, Hansen HH (1982) Lung cancer. In: Holland JF, Frei E III (eds) Cancer medicine, 2nd edn. Lea and Febiger, Philadelphia, pp 1709–1752

Snow JH, Goldstern HM, Wallace S (1979) Comparison of scintigraphy, sonography and computed tomography in the evaluation of hepatic neoplasms. Am J Radiol 132: 915

Sommer B, Bauer WM, Rath M, Frenzel G, Stelter WJ, Lissner J (1981) Die computertomographische Stadieneinteilung des Bronchialcarcinoms. Auswirkungen auf das diagnostische und therapeutische Vorgehen. Computertomographie 1: 131–135

UICC (1978) TNM Classification of malignant tumours, 3rd edn. UICC, Geneva

Vas W, Zylak CJ, Mather D, Figeredo (1981) The value of abdominal computed tomography in the pre-treatment assessment of small cell carcinoma of the lung. Radiology 138: 417–418

Vock P, Haertel H (1981) Die Computertomographie zur Stadieneinteilung des Bronchuscarcinoms. Fortschr Röntgenstr 134: 131–125

Wallace S, Chuang VP (1982) The radiologic diagnosis and management of hepatic metastases. Radiologe 22: 56–64

Webb WR, Gamsu G, Glazer G (1981) Computed tomography of the abnormal pulmonary hilum. J Comput Assist Tomogr 5: 485–490

WHO Handbook for Reporting Results of Cancer Treatment (1979) World Health Organization, Geneva, publication no 88

Wilson MA, Calhoun FW (1981) The distribution of skeletal metastases in breast and pulmonary cancer: concise communication. J Nucl Med 22: 594–597

Wouters EFM, Oei TK, van Engelshoven JMA, Lemmens HAJ, Greve LH (1982) Evaluation of the contribution of computed tomography to the staging of non-oat-cell primary bronchogenic carcinoma. A retrospective study. Fortschr Röntgenstr 137: 540–543

Zelen M (1973) Keynote address on biostatistics and data retrieval. Cancer Chemother Rep Part 3, 4: 31–42

The Role of Surgery in the Treatment
of Small Cell Carcinoma of the Lung

W. Maassen, D. Greschuchna, and I. Martinez

Ruhrlandklinik, Tüschener Weg 40, 4300 Essen 16, Federal Republic of Germany

Introduction

While the importance of *resection/surgery* in the treatment of carcinoma of the lung is fully acknowledged, *surgical therapy in the presence of a small cell carcinoma of the lung* is vigorously disputed. There seem to be various reasons for this dispute.

1) *The rejection on principle of any sort of operative treatment* is primarily based on a study by Fox and Scadding (1973) for the British Medical Research Council. In a comparison of 71 patients treated by surgery and 73 treated by radiotherapy, in which the method of treatment was randomly assigned, 4 of the 73 treated by radiotherapy survived 5 years. Of the patients in the surgical group, 1 survived; he had refused operation and lived for 8 years following radiotherapy.
A study by Takita et al. (1973), covering a 12-year period, 1960–1971, and involving 161 patients with small cell carcinoma, also showed that the average survival time was only 4.6 months and surgery was not curative for any patient.
In 1969, Miller et al. emphasized the importance of radiotherapy as the method of choice in the treatment of small cell carcinoma of the lung, as a result of a comparative study he had carried out involving 144 patients. He found average survival periods of 199 days following surgery and 284 days after radiotherapeutic treatment, with 2-year survival rates of 3% and 7% respectively, and 5-year rates of 1% and 4%, respectively.

2) *Widely varying results concerning survival rates after the surgical treatment of small cell carcinoma of the lung* appear in the literature: between 0 and 35% (Broder et al. 1977; Buchberger et al. 1979; Freise et al. 1978; Greschuchna and Maassen 1975, 1980; Greschuchna 1978; Higgins et al. 1975; Jenny 1971; Konrad et al. 1980; Lennox et al. 1969; Martini et al. 1975; Pichlmaier and Junginger 1974; Rostad and Lexon 1979; Shore and Paneth 1980; Widow 1973).
Of course, in the light of this divergency, it is essential to ask whether different criteria for selection were used (the stages of the tumors are often not stated) and whether the histological findings always confirmed small cell carcinoma beyond doubt.

3) Furthermore, *the different experiences of the various authors in terms of diagnostic and surgical therapy* also play a role. Junginger (1980) very much doubts that preoperative cytological or histological evidence of a small cell carcinoma without proof of metastases is reason enough to exclude peripheral small cell carcinoma from operative treatment. Our experience has shown his point to be justified. Of 25 tumors revealed as small cell carcinoma after operation, 4 had been misdiagnosed as squamous cell carcinomas preoperative, 2 as large cell and 1 as adenocarcinoma. The preoperative diagnosis of central small cell carcinoma of the lung was found by resection to be correct in 90% of

Recent Results in Cancer Research. Vol. 97
© Springer-Verlag Berlin · Heidelberg 1985

cases, but was confirmed in only 17% of cases of peripheral carcinoma. Typing according to cytological tests on bronchial secretion and findings after operation also showed considerable differences: of 9 small cell carcinomas, 4 had previously been misdiagnosed as squamous cell carcinomas and only 5 had been correctly diagnosed preoperatively.

In central and peripheral small cell carcinomas we found identification rates of 40% and 75%, respectively, with preoperative cytological typing later compared with the resected specimen. The false diagnosis rate for central small cell carcinoma of the lung was particularly striking: only 2 of 5 tumors had been correctly diagnosed, 3 having been identified as squamous cell carcinomas (Greschuchna et al. 1983).

Shore and Paneth (1980) only considered small cell carcinomas as inoperable if bronchoscopic examination showed them to be such, distant metastases from the primary tumor were present, or the poor general condition of the patient made surgery inadvisable. From 1959 to 1974 they achieved a 5-year survival rate of 25% and considered surgical treatment of small cell carcinomas of the lung to be a reasonable and promising method of treatment in suitable cases. Broder et al. (1977) and Clifton (1966) also disagree with Mountain's (1974) recommendation that small cell carcinomas of the lung should be excluded from any sort of resection. As early as 1968, Lennox et al. (1969), in a retrospective study of 275 patients operated on for small cell carcinoma of the lung, reported a 2-year survival rate of 19% following resection; the prognosis was dependent on the extent of the tumor and, in particular, on the stage of lymphogenic spread. When no lymph node metastases were present, 28% of their patients survived 2 years after surgery, whereas only 11% of those with such metastases at the time of operation lived so long postoperatively.

4) After others authors (Krish et al. 1976; Levison 1980; Paulson and Reisch 1976; Bennet and Smith 1978; Rubinstein et al. 1979; Naruke et al. 1976; Rostad et al. 1979) had also stated a preference for surgery over any other form of treatment, further discussion was enlivened by *three recent studies*. The findings of Higgins et al. (1975), in a paper for the Veterans Administration Armed Forces Study, should be mentioned. These authors found that the survival rate of patients with solitary round foci at the same tumor stage was *identical regardless of the type of resected carcinoma*. Four of eleven patients who had had a small cell carcinoma survived for 5 years (36%), and two survived for 10 years (18%). This corresponds with our own results (Greschuchna 1978; Greschuchna and Maassen 1980).

Meyer et al. first reported on their experience and point of view in 1979, but in 1982 they changed their attitude towards the operative treatment of small cell carcinomas of the lung, giving various reasons for this. On the one hand, in cases treated with combined radiotherapy and chemotherapy local recurrence of small cell carcinomas was the most common, and therefore it was recommended that any accessible tumor should be excised. On the other, radiotherapy as an alternative to surgical treatment would limit the application of chemotherapy. Their main argument was that the complete *extirpation of a small cell carcinoma of the lung is equivalent to a total remission* without the after-effects of an immunosuppressive therapy. Chemotherapy can therefore be more intensively and effectively applied. In further studies, the authors recognized the necessity for a more exact TNM classification, so that the benefit of surgical treatment could be more precisely noted and evaluated. The group reports on only ten treated patients in stages I and II of the disease with initial resection. One patient died of a pulmonary embolism postoperatively, while nine were treated with adjuvant chemotherapy and remained free of recurrences for between 7 and 69 months. Of six patients with stage III tumors, five had no recurrence 25 months later; one recurrence appeared 26 months postoperatively. Four of the ten patients treated with initial resection had T2N1 tumors, four T2N0, and the remaining two had tumors classified as T1N0. Eight lobectomies were performed, one pneumonectomy and one segment resection. The authors regard a complete extirpation of tumors, even in stage III, as helpful for chemotherapeutic treatment, although they see surgery as contraindicated when the tumor is known to be T3, or N2 preoperatively. They point out that a rapid reaction to initial chemotherapy can also be a clear indication for surgery. Of course, this must then be performed within a short time, because with combined radiotherapy and chemotherapy recurrences appear, even in cases of limited disease, from the fourth month on. If chemotherapy does not produce any clear reaction resection is not indicated. They further point out that cerebral radiotherapy is just as necessary with surgery as with radiotherapy/chemotherapy alone. When lung resection is performed as a secondary treatment following initial chemotherapy, the whole of the area that was affected *before* the initial therapy should be removed. They see other grounds for exclusion from surgery in involvement of the bifurcation, superior vena cava syndrome, contralateral involvement, malignant pleurisy, poor pulmonary function. limited reaction to chemotherapy, as already mentioned, of course, refusal of the operation by the patient.

Mayer et al. (1982) found that small cell carcinomas of the lung with polygonal cells have a better prognosis than those with nonpolygonal cells. It has not yet been verified in the literature whether these findings are significant.

Shields et al. (1982) and Meyer et al. (1982) report that better survival rates are achieved with patients receiving a combined treatment with surgery and chemotherapy, particularly in cases of limited disease, even when the disease is in stage T2N2. With chemotherapy alone, clinical remission is achieved in 50% of the patients, but the recurrence rate is 70%.

The *largest current study* is that of Shields et al. (1982), in material collected from 25 clinics of the US Veterans Administration. In all, 148 patients underwent surgery for small cell carcinomas of the lung. The postoperative death rate was high, and the 16 patients concerned were excluded from further consideration and calculation of the statistics, because they did not fit into either of the branches of chemotherapy alone, or radiotherapy and chemotherapy in combination. The authors reported an overall 5-year survival rate of 23%, calculated using the so-called life-table method. However, if we take the absolute numbers, only 23 out of 148 patients survived as long as 5 years, wich is 15.5%. The appropriateness of the life-table method for a study of the immediate results of surgery for small cell carcinoma within a 3- or 5-year period also seems questionable. In view of the cardiorespiratory functional condition necessary for these operative procedures, these patients show rather greater lif expectancy than that of other people of the same age. For this reason, although the use of such a method is statistically allowed, it is misleading and questionable in terms of the differences given. We have not used this method in our studies, and the postoperative lethality is therefore included in our results.

Shields et al. (1982) give the following 5-year survival rates for the various tumor stages: 60% for T1N0M0, 31% for T1N1M0, 28% for T2N0M0, and 9% for T2N1M0. The survival rate was only 3.6% when either T3 or N2 was present (numbers according to the life-table method).

This is the only recent study in which the influences of adjuvant chemotherapy in association with or following surgery has been evaluated. The resulting small patient groups must be taken into account when its conclusions are considered. In the calculations in this study, too, the authors use the life-table method. In the early trials, adjuvant chemotherapy consisted the use of nitrogen mustard of cyclophosphamide, and in the later trials, with a considerably longer regimen, cyclophosphamide alone, cyclophosphamide alternating with methotrexate and, finally, CCNU and hydroxyurea. A 5-year survival rate of 16.3% is calculated for the early period, and 34.7% for the later one; the final calculation for the two together, als already mentioned, is then 23%. In terms of tumor stages, the 5-year survival rate of T1N0M0 was 59.5%, for T1N1M0 31.3%, for T2N0M0 27.9%, for T2N1M0 9%, and T3 or N2 3.6%; these percentages are all calculated according to the life-table method, postoperative deaths having been excluded.

Surprising results of treatment were given for the best chemotherapy group, CCNU and hydroxyurea, with 81% as against 38% in the control group.

The authors consider that T1N0M0, and probably also T1N1M0 or T2N0M0, constitute definite indications for resection. In other stages they see a contraindication to surgery. The stage T1N0M0 was the only one to show statistically significant differences from the other stages seen as possibly suitable for surgery.

In this respect it seems rather surprising that, in the latest review in *Thoracic Oncology*, edited by Choi and Grillo (1983), no importance is attributed to small cell carcinomas in the section on surgical indication. This is also the case in the survey by Hansen (1982).

Overall it must be stated that, especially recently, *a change in concept concerning the surgical treatment of small cell carcinoma of the lung* can be seen. All the authors who have addressed themselves to this particular problem are agreed that selective surgery should be seen as a component of an integrated therapy for small cell carcinoma of the lung, as we stated earlier, and on no account as an alternative to chemotherapy/radiotherapy alone. The possibility of surgical treatment is more limited in these tumors than in the other differentiated forms of carcinoma of the lung, because of its tendency to spread quickly.

Personal Experience 1962–1979

As shown in Table 1, a total of 109 patients with small cell carcinoma of the lung were operated on in our clinic from 1962 to 1979. The resection rate of 94 out of 109 (86%) shows that the percentage of exploratory thoracotomies in cases with previous negative mediastinoscopy is considerably higher for this type of tumor than for carcinomas of the

Table 1. Surgical therapy of small cell carcinoma (1962–1979)

Operation performed	n	%
Thoracotomy	109	100
Exploratory	15	14
Resections	94/109	86
Extended pneumonectomy	25/ 94	27
Pneumonectomy	38/ 94	40
Lobectomy	28/ 94	30
Segmental resection	3/ 94	3
3 year survival	21/ 94	22.3

lung in general. Previously the figure for the latter was 11%, and more recently between 3% and 5% of our patient population.

We have presented the 3-year survival rate because this is the only one that allows comparison with the results of combined radiotherapy and chemotherapy. In these statistics the 3-year survival rate is usually given.

Table 1 also shows that pneumonectomy and extended pneumonectomy are more frequent for small cell carcinoma of the lung than for other malignant lung diseases; segment

Table 2. Surgical therapy of small cell carcinoma (1962–1979)

	1962–1975			1976–1979		
	T	R	3-Y.S.	T	R	3-Y.S.
Stage I						
T1 NO MO	12	12	7/12 = 58%	2	2	0/ 2
T1 N1 MO	7	7	1/ 7 = 14%	2	2	0/ 2
T2 NO MO	6	6	2/ 6 = 33%	5	5	1/ 5 = 20%
Stage II						
T2 N1 MO	11	11	1/11 = 9%	5	5	3/ 5 = 60%
Stage III						
T1 N2 MO	3	3	2/ 3 = 67%	3	2	1/ 2 = 50%
T2 N2 MO	6	5	0/ 5	6	3	1/ 3 = 33%
T3 NO MO	2	2	0/ 2	2	2	0/ 2
T3 N1 MO	5	5	0/ 4	5	5	1/ 5 = 20%
T3 N2 MO	21	15	1/15 = 7%	5	2	0/ 2
Total	73	66 = 90%	14/65 = 22% 1?	36	28 = 78%	7/28 = 25%

	1962–1979			?	3-year survival by stage
	T	R	3-Y.S.		
Stage I					
T1 NO MO	14	14	7/14 = 50%	–	
T1 N1 MO	9	9	1/ 9 = 11%		11/34 = 32%
T2 NO MO	11	11	3/11 = 27%		
Stage II					
T2 N1 MO	16	16	4/16 = 25%	–	4/16 = 25%
Stage III					
T1 N2 MO	6	5	3/ 5 = 60%	–	
T2 N2 MO	12	8	1/ 8 = 13%	–	10/59 = 17%
T3 NO MO	4	4	0/ 4 = 0%	–	6/43 = 14%
T3 N1 MO	10	10	1/ 9 = 11%	1	
T3 N2 MO	27	17	1/17 = 6%	–	
Total	109	94 = 86%	21/93 = 23%	1	21/93 = 23% 21/94 = 22%

T, thoracotomy; R, resection; 3-Y.S., 3-years survival

resection is not very common in these cases. Of the 94 patients who underwent resection, 93 (99%) were followed-up.

The 3-year survival rate of 21 of 94 patients (22.4%) in absolute numbers is calculated, with due regard for postoperative lethality and the varying periods of chemotherapy. This rate is similar to Shields et al.'s (1982) results, which were calculated *with exclusion of postoperative lethality* and *with figures extrapolated from random values (life-table method)*.

Table 2 gives the 3-year survival rates for the periods 1962–1975, then 1976–1979, and finally, combined figures for the whole period 1962–1979. These results show that there was no significant difference between the survival rate in the first period (14/65, 22%) and that in the second (7/27, 26%). A considerably higher overall rate of 32.4% is seen for stage I tumors than for the other tumor stages, in which the survival rate drops drastically to 17%: for stage II only 25% and for stage II 14%.

It can also be noted from Table 2 that considerably more early manifestations, represented particularly well in stage I tumors, were surgically treated in the first period than in the second. The resections undertaken from 1976 to 1979 were concerned with stage II and III tumors to a much greater extent. In the light of this fact, it is even more surprising that the results do not vary much from those of the first period. These results lead to the following conclusions:

1) The earlier the initial operation on small cell carcinoma of the lung is, the greater are the chances of survival for 3 or 5 years with adjuvant radiotherapy/chemotherapy.

2) The present forms of adjuvant (radiotherapy)/chemotherapy also seem to lead to better survival rates – even for advanced forms of small cell carcinoma of the lung, and especially even in cases of lymph node stage N2. Providing mediastinoscopy is negative and the metastatic stage is M0.

3) With the strictly localized forms of small cell carcinoma, i.e., particularly those in stage I, but also in stage II, resection is, without doubt, important in the treatment schedule; in our opinion it is essential.

4) Even in stage III with more extensive disease, although limited to the ipsilateral hemithorax, the combination of resection with chemotherapy and/or radiotherapy/chemotherapy is certainly of value in view of the extent of a particular small cell tumor. This becomes clear when one considers that nearly all the patients in the second period, from 1975 to 1979, who survived 3 years had tumors in stages T1N2 to T3N2. These observations, incidentally, contradict all the efforts at early diagnosis of carcinomas of the lung in the decades since 1960. It should be pointed out here that unfortunately it was not possible for us to examine the various forms of adjuvant (radiotherapy)/chemotherapy in connection with the survival rates, as the patients who are operated on in our clinic come from a wide area, so that preoperative and postoperative treatment cannot be verified in detail.

Table 3. Three-year survival correlated with isolated N and T stages[a]

Stage	*n*	%	Stage	*n*	%
N0	10/29	35	T1	11/28	39
N1	6/34	18	T2	8/35	23
N2	5/30	17	T3	2/30	7
N1 + 2	11/64	17	T2 + 3	9/64	14

[a] Stage I, 11/34 = 32%; Stage II, 4/16 = 25%; Stage III, 6/43 = 14%

The experiences of various authors in the literature and our own results (Tables 2 and 3) lead us to the following conclusions:

1) Surgical treatment of small cell carcinoma of the lung can no longer be seen as obsolete when the uncertainty of the preoperative diagnosis in terms of classification of the tumors on presentation, especially, is taken into account.

2) Obviously the decision to operate presupposes intensive preoperative staging, and this should include not only bronchoscopy to establish the extent of the tumor itself; preoperative mediastinoscopy (60% show mediastinal metastases), bone scintigram, total-body computed tomography, epigastric sonography, and iliac crest biopsy also seem indispensable to us.

3) The surgical treatment can only be a part of an integrated therapy including the possibility of (radiotherapy)/chemotherapy.

4) With reference to the optimal time for surgery, at present it can only be said that an initial resection is both possible and necessary for tumors in stage T1N0 and T2N0. It is questionable whether the same applies for stage T2N1, but this cannot usually be decided preoperatively.

5) Tumors in stage III are not normally resected initially, but at most after preliminary chemotherapy. The question arises here as to whether only those cases that have responded quickly should be treated surgically, or also those that have manifested no reaction to chemotherapy according to clinical and radiological findings. This decision should probably favor the quick responders rather than the patients whose tumors show no appropriate response.

Summary

The role of surgical treatment of non-small-cell carcinoma of the lung is controversial. Surgical therapy of small cell carcinoma of the lung has been the subject of criticism for two decades — in contrast to our opinion about the important role of initial surgical therapy in limited disease of this type of lung carcinoma. In a review of the results of surgical therapy in 109 patients with undifferentiated small cell carcinoma of the lung in the period from 1962 to 1979 and an attempt to define the role of the curative effect, we report on 109 thoracotomies after negative preoperative mediastinoscopy and exclusion of hematogenous spread of the cancer. The resection rate (94/109) was considerably lower than in differentiated lung carcinoma. Pneumonectomy (40%) and extended pneumonectomy (27%) were more frequently performed for small cell carcinoma than for differentiated tumors; thus, the rate of lobectomy and segmental resection (28/94 = 30%, and 3/94 = 3%, respectively) was low. When we included the postoperative lethality in the calculation rather than using the life-table method, we found in a follow-up study of 99% of the patients who had undergone resection, in absolute numbers, 3-year survival in 21/93 patients (23%). In the different stages I, II, and III, we noted 3-year survival rates of 32%, 25%, and 14%. Correlation to the N and T stages was N0 (10/29) 35%; N1 (6/34) 18%; N2 (5/30), 17%; N1 + 2 (11/64) 17%; T1 (11/28) 39%; T2 (8/35) 23%; T3 (2/30) 7%; and T2+3 (9/34) 14%.

We conclude that initial surgical therapy is as important in an integrated approach to therapy including radio-/chemotherapy for stages I and II as for selected cases in stage III.

References

Bennet WF, Smith RA (1978) A twenty-year analysis of the results of sleeve resection for primary bronchogenic carcinoma. J Thorac Cardiovasc Surg 76: 840–845

Broder LE, Cohen MH, Selawry OS (1977) Treatment of bronchogenic carcinoma. II. Small cell. Cancer Treat Rev 4: 219

Buchberger R, Jenny H, Strahberger E (1979) Drei Jahrzehnte Resektionsbehandlung beim Bronchuskarzinom. Wien Klin Wochenschr 91: 101

Choi NC, Grillo HC (1983) Thoracic oncology. Raven, New York

Clifton EE (1966) The criteria for operability and resectability in lung cancer. JAMA 195: 1031

Fox W, Scadding JG (1973) Medical research council comparative trial of surgery and radiotherapy for primary treatment of small-celled or oat-celled carcinoma of bronchus. Lancet 2: 63–65

Freise G, Gabler A, Liebig S (1978) Bronchial carcinoma and longterm survival. Thorax 33: 228

Greschuchna D (1978) Ergebnisse der operativen Behandlung des kleinzelligen Bronchialkarzinoms. Thoraxchirurgie 26: 300–303

Greschuchna D, Maassen W (1975) Ergebnisse der sogenannten erweiterten Resektionen beim präoperativ routinemäßig mediastinoskopisch untersuchten Bronchialkarzinom. Kongr Ber Wiss Tagg Norddtsch Ges Lungen Bronchialheilkd 14: 100

Greschuchna D, Maassen W (1980) The importance of histological classification and tumor staging for prognosis after resection of bronchial carcinoma. Thorac Cardiovasc Surg 28: 115

Greschuchna D, Maassen W (1981) Pro-Contra: Kleinzelliges Bronchialkarzinom – Operation oder Chemotherapie als primäre Behandlung. Internist (Berlin) 22

Greschuchna D, Kasparek R, Kappes R (1983) Ein Vergleich der Histologien präoperativer Bronchusbiopsien und Mediastinalbiopsien mit Lungenresektaten von Bronchialkarzinomen. Prax Klin Pneumol 37: 862–865

Hansen HH (1982) Management of small cell anaplastic carcinoma 1980–1982. In: Ishikawa S, Hayata Y, Suemasu K (eds) Lung cancer 1982. Excerpta Medica, Amsterdam

Higgins GA, Shields TW, Keen RJ (1975) The solitary pulmonary nodule. Arch Surg 110: 570

Jenny H (1971) Histologie, Tumorstadium, Lebenserwartung beim Bronchialkarzinom. Thoraxchirurgie 19: 244

Junginger T (1980) Indikation zur Operation bei Bronchialkarzinom. Dtsch Med Wochenschr 105: 674

Kirsh MM, Rotman H, Argenta L, Bove E, Cimmino V, Tshian J, Ferquson P, Sloan H (1976) Carcinoma of the lung: Results of treatment over ten years. Ann Thorac Surg 21: 371–377

Konrad RM, Ammedick U, Bläute R (1980) Soll das kleinzellige Bronchialkarzinom operiert werden? Med Welt 31: 1097

Lennox SC, Flavell G, Pollok DJ, Thompson VC, Wilkins JL (1969) Results of resection for oat-cell carcinoma of the lung. Lancet 2: 925

Levison H (1980) What is the best treatment for early operable small cell carcinoma of the bronchus? Thorax 35: 721–723

Martini N, Wittes RE, Hilaris BS (1975) Oat cell carcinoma of the lung. Clin Bull 5: 144

Mayer JE, Ewing SL, Ophoven JJ, Summer HW, Humphrey EW (1982) Influence of histologic type on survival after curative resection for undifferentiated lung cancer. J Thorac Cardiovasc Surg 84: 641–648

Meyer JA, Comis RL, Ginsberg SJ, Ikins PM, Burke WA, Parker FB (1979) Selective surgical resection in small cell carcinoma of the lung. J Thorac Cardiovasc Surg 77: 243–248

Meyer JA, Comis RL, Ginsberg SJ, Pkins PM, Burke WA, King GA, Gullo JJ, DiFino SM, Tinsely RW, Parker FB (1982) Phase II trial of extend indications for resection in small cell carcinoma of the lung. J Thorac Cardiovasc Surg 83: 12–19

Miller AB, Fox W, Tall R (1969) Five-year follow-up of the medical research council comparative trial fo surgery and radiotherapy for the primary treatment of small celled or oat-celled carcinoma of the bronchus. Lancet 2: 501–505

Mountain C (1974) Surgical therapy in lung cancer: Biologic, physiologic, and technical determinants. Semin Oncol 1: 253

Naruke T, Suemasu K, Ishikawa S (1976) Surgical treatment for lung cancer with metastasis to mediastinal lymph nodes. J Thorac Cardiovasc Surg 71: 279–285

Paulson DL, Reisch JS (1976) Long-term survival after resection for bronchogenic carcinoma. Ann Surg 184: 324–332

Pichlmaier H, Junginger T (1974) Diagnostik und Therapie des Bronchialkarzinoms. Münch Med Wochenschr 116: 137

Rostad JR, Lexon VP (1979) Survial in lung cancer after surgery. Scand J Respir Dis 60: 297

Rostad H, Vale JR, Lexow P (1979) Survival in lung cancer after surgery. Scand J Respir Dis 60: 297–302

Rubinstein I, Baum GL, Kalter Y, Pauzner Y, Liebermann Y, Bubis JJ (1979) The influence of cell type and lymph node metastases on survival of patients with carcinoma of the lung undergoing thoracotomy. Am Rev Respir Dis 119: 263–262

Shields TW, Ill C, Higgins GA, Matthews MJ, Keehn RJ (1982) Surgical resection in the management of small cell carcinoma of the lung. J Thorac Cardiovasc Surg 84: 481–488

Shore DF, Paneth M (1980) Survival after resection of small cell carcinoma of the bronchus. Thorax 35: 819

Takita H, Brugarolas A, Marabellea P, Vincent RG (1973) Small cell carcinoma of the lung. Clinicopathologic studies. J Thorac Carciovasc Surg 66: 472–477

Widow W (1973) Die Bedeutung von jährlichen Röntgenreihenuntersuchungen für die Erfassung und Behandlung des Bronchialkarzinoms. Dtsch Gesundheitswes 28: 2410

The Role of Radiotherapy in the Management of Small Cell Bronchogenic Carcinoma

N. M. Bleehen and D. H. Jones

University Department of Clinical Oncology and Radiotherapeutics, The Medical School, Hills Road, Cambridge CB2 2QH, England

Introduction

Following its discovery in 1895, radiation was used therapeutically in an attempt to control lung cancer. Unfortunately, success was tempered by the fact that the energy of the radiation was inadequate, which also resulted in cosiderable complications, especially if operative intervention followed. With the advent of megavoltage machines, more and more successes were reported in lung cancer generally, and the controversy arose as to whether radiation should be used alone or in conjunction with surgery. However, it became quickly apparent that such a controversy had far more ramifications when applied to small cell lung cancer, mainly because of the unrecognised degree of spread of the disease at presentation. Despite dramatic improvements in investigative techniques over the last two decades, the successful treatment of small cell lung cancer is considerably limited by its early occult spread.

Radiotherapy may be applied in a variety of ways in small cell lung cancer. Its palliative role is immediately obvious in that it can be of considerable benefit in providing symptomatic relief. Its curative or radical role is not so easily apparent, especially as its value in association with other modalities is less easy to assess. Its spectrum of use extends from its administration solely for locoregional disease to treatment of the whole body, and in association with chemotherapeutic agents to prophylactic irradiation. Discussions of the problems of X-ray therapy in the treatment of small cell lung carcinoma may be found in recent review articles by Bleehen (1979), Salazar and Creech (1980), and Bleehen et al. (1983). Other short reviews have been published by Byhardt and Cox (1983) and Cohen (1983), and these indicate the difficulty in producing a unified approach to the use of radiotherapy in small cell lung cancer.

Radiotherapy as a Curative Treatment for Limited Disease

Preoperative Radiotherapy

It is inevitable that patients undergoing surgery for small cell lung cancer from a highly selected group. Mouritzen et al. (1982) suggest that surgery alone can produce a 5-year survival figure of 10%−15% in this disease (figures based on 146 patients over 27 years) and that better results must be obtained if radiation is to be given in association with surgery. Levison (1980, 1982) quotes a 4-year survival of 17% in patients given 1,750 cGy (in 7 fractions) preoperatively, but nil 4-year survival in a similar group of patients (nonrandomised studies) given 2,500 cGy (in 10 fractions). Other authors (Higgins et al.

Recent Results in Cancer Research. Vol. 97
© Springer-Verlag Berlin · Heidelberg 1985

1982; Karrer et al. 1982; Nakagawa et al. 1982; Ohta et al. 1982; Daddi et al. 1982) report 3- to 5-year survival rates between 16% and 19% in patients treated surgically, with or without adjuvant chemotherapy. All these recent studies together suggest that, for *early* disease, surgery alone can produce an acceptable 5-year survival figure, and that this may not necessarily be improved by the addition of another modality. However, there is no randomised large-scale study available to assess whether surgery with preoperative radiotherapy has any benefit over surgery alone.

Local Radiotherapy Alone. The question arising from the discussion above is: how does radiotherapy alone compare with surgery alone? The United Kingdom Medical Research Council (MRC) approached this problem. Their report (Miller et al. 1969) on the MRC first small cell study showed that in 144 patients randomly allocated to surgery or radical radiotherapy the 4-year survivals were 3% and 7%, respectively. The mean survival times were 7 and 10 months, respectively, the difference being statistically significant at $P = 0.05$. The 10-year update (Fox and Scadding 1973) also indicated improved median survival time (at 11 months) ($P = 0.04$) for the radiotherapy-treated group. This rather firm illustration of the benefit of radiotherapy over surgery has been questioned recently, and certainly the 5-year survival figures of between 12% and 16% quoted above for surgical management of early small cell cancer are far better than those reported for either modality in the MRC study. Of course, numerous factors will determine the quoted results: accurate staging (especially with the more sophisticated techniques used over the last 10 years); extent of surgery performed; radiation doses; fractionation and treatment volume; initial patient performance status; randomisation of studies, etc. Hence it may be unfair to compare the results of these studies directly, but such a comparison does illustrate that in early disease, small cell lung cancer may be cured by operative intervention.
Small cell lung cancer, when untreated, has a median survival of 12 weeks for limited disease and 5 weeks for extensive disease (Roswit et al. 1968; Wolf et al. 1966). The MRC study (Fox and Scadding 1973) has shown that radical radiation results in a median survival time of 11 months in limited disease. So, for patients with limited but inoperable disease, radical radiotherapy was the mainstay of treatment prior to the advent of chemotherapy. The treatment volume normally includes the site of primary disease together with the regional lymph nodes, which may contain occult disease. Often such a plan results in a large volume of about 10 × 15 cm cross section, so that tolerance of surrounding normal tissues and the total dose given must be carefully considered. There is no unequivocal evidence that the use of shrinking fields produces better results (or fewer complications) than fixed fields. Whether the spinal cord and oesophagus need shielding from the radiation, and if so, at what point in treatment this should be done, also remains uncertain and currently this is left to the individual therapist's practice.
Radiation dose and fractionation vary considerably between centres. Salazar and Creech (1980) quote a dose of 40−50 Gy (> 1,400 ret) with conventional fractionation as necessary to achieve total control of the primary, when the treatment is given in conjunction with chemotherapy. Rubin et al. (1976) and Perez (1977) suggest doses in excess of 50 Gy in 5 weeks when radiation is given as a single modality. However, Deeley (1966) reports that "for inoperable anaplastic carcinoma of bronchus" 30 Gy in 20 fractions over 4 weeks is better than 40 Gy over the same period (but no lung correction was applied in this trial). In a small number of patients, Choi and Carey (1976) described a dose-response relationship with 30 and 50 Gy, in that there was better immediate local control with the higher doses. Fewer recurrences were demonstrated by Cox and his co-workers (1978) with a median dose of around 1,700 ret than with one of 1,450 ret. It is

therefore difficult to deduce whether there is a dose-dependent response, and it is not possible to propose a single value as being an optimal dose. When radiation is given in conjunction with chemotherapy, the combined effect on the primary tumour and the local tissues almost certainly means that the higher radiation doses cannot be used, and Cox et al. (1978) showed that adequate control of the tumour was possible with a lower median dose (about 1,300 ret) in the presence of chemotherapy. It was for this reason that the MRC second (1979, 1981) and third (1983) small cell cancer studies selected 30 Gy in 15 fractions over 3 weeks as the dose to be used in conjunction with a chemotherapy regimen. The present fourth MRC trial however uses a higher dose of 40 Gy in 15 fractions in 3 weeks, and it will be interesting to note when this trial is reported whether there is a greater toxicity from the higher dose of radiation. There is no evidence to suggest that a split course of radiation gives better or worse tumour control than a standard continuous course.

Local Radiotherapy in Conjunction with Adjuvant Chemotherapy. Because of the propensity of small cell lung cancer to metastasise early and in an occult manner, it was natural that the addition of chemotherapeutic agents to the already established radiotherapeutic method of treatment came to form the basis of many new trials. One of the largest studies performed using randomisation of patients to receive radiotherapy or radiotherapy and chemotherapy was that of the MRC. The results of this study (Medical Research Council Lung Working Party 1979, 1981) in 236 patients receiving 30 Gy in 15 fractions over 3 weeks with or without a three-drug chemotherapy regimen (cyclophosphamide, methotrexate, and CCNU) showed a significant increase in median survival time of the patients who received chemotherapy. The median survival times for this study, together with the results of other but smaller randomised studies, are shown in Table 1. In these studies, intrathoracic tumour recurrence was a major cause of relapse, occurring in some 30% of patients. The median survival time frequently lacks precision, especially if the numbers of patients in a particular study are small. However, it is a parameter that is used in many trials because very often no others are quoted. Improved accuracy of reporting would be obtained from crude and actuarial survival times at or beyond 2 years, and in future studies these must be recognised as being more valuable than median survival.
Encouragingly, further studies have shown an improvement in median survival — especially with new combinations of drugs. The nonrandomised trials of Livingston et al. (1978), Holoye et al. (1977, 1978, 1983), Einhorn et al. (1978), Greco et al. (1979), and

Table 1. Randomised trials of radiotherapy versus radiotherapy with chemotherapy

Investigator	Number of patients	Median survival (months)	
		RT	RT + CT
Bergsagel et al. 1972	41	5	9.5
Høst 1973	75	5.5	8.5
Tucker et al. 1973	22	3	6[a]
MRC 1979	236	6	9[a]
Krauss and Perez 1979	61	5	10[a]
Petrovich et al. 1977	59	5	9.5[a]
Matthiessen 1978	34	5	12[a]

[a] $P = {} < 0.05$

Johnson et al. (1978) collectively show a median survival ranging from 12 to 18 months, the average response rate being about 75%. With such an improvment in median survival time, some centres have further investigated the possibility of completely excluding radiotherapy in the management of small cell lung cancer, except for palliation or prophylactic cranial irradiation.

In their retrospective studies. Salazar and Creech (1980) report a considerably greater intrathoracic relapse rate in patients receiving chemotherapy alone, as compared with chemotherapy and radiotherapy together (82% and 28%, respectively). However, this study was retrospective and nonrandomised, with gross imbalance of patient numbers. Another retrospective study, by Bunn and Ihde (1981), yielded a better 2-year survival rate for the combined treatments (17%) than for chemotherapy alone (7%).

More recent randomised studies investigating whether the addition of radiotherapy to chemotherapy is of value will hopefully provide a more definite answer. The results of Williams et al. (1977) and Stevens et al. (1979), together with the larger study of Dombernowsky et al. (1980), seem to suggest a marginally better median survival time of 11−12 months for chemotherapy alone, compared with 9−11.5 months for chemotherapy plus radiotherapy. Direct comparison of all these studies may not be justified as the extent of disease was different, prophylactic cranial irradiation was given in some but not others, and the radiation dose to the chest also varied. Dombernowsky et al. reported a much greater locoregional recurrence rate (78%) in the absence of radiotherapy (48% if radiotherapy was given), but the rather worse median survival time in the combined modality group may reflect what some clinicians would consider rather intensive radio-therapy.

Preliminary reports of the studies of Fox et al. (1980) and Cohen et al. (1980) have been published, and give identical median survival times of around 14 months for chemotherapy and for combined radio- and chemotherapy. Harper et al. (1983), in a large randomised study of 366 patients, have suggested that patients receiving radiotherapy + chemotherapy are half as likely to relapse at the primary site, but to date radiotherapy does not confer a survival advantage. The preliminary report published by Perez et al. (1983) of a study in over 300 patients indicates a higher response rate (84% and 66%), a greater median survival time (17 months and 12 months), a much higher 3-year survival rate (20% and 5%; $P = 0.02$), and a lower intrathoracic relapse rate (33% and 48%) in patients receiving radiotherapy plus chemotherapy and chemotherapy alone, respectively (both groups receiving whole-brain irradiation), but no difference in the incidence of distant metastasis.

The studies of Souhami et al. (1983) suggest that intrathoracic radiation may be given in conjunction with high-dose cyclophsophamide (160−200 mg/kg) and autologous marrow transplantation. In their series of 25 patients 21 had limited disease. The overall median survival was 17 months, and only one patient died as a direct result of treatment. Despite such an energetic approach in association with radiation to the primary site, relapse still occurred in most patients.

As the disease is often metastatic at presentation, it seems logical to commence treatment with chemotherapy and then consolidate the intrathoracic response with radiation. It is difficult to know, however, whether the radiation treatment volume should be calculated for the now reduced tumour mass or for the original tumour mass (especially in view of the high frequency of recurrence at the original site). Commencing treatment with radiation to bulky disease has the advantage of giving treatment to the main primary mass initially, but this could be at the expense of continued metastatic growth. The third MRC Small Cell Study (Medical Research Council Lung Cancer Working Party 1983) of 91 patients randomised to receive either radiotherapy to the primary site, followed by chemotherapy, or chemotherapy, followed by radiotherapy, and then further chemotherapy, has shown a nonsignificant difference in median survival times of 9 and 11 months, respectively. The 3-year survival values are 8% and 1%, respectively. It seems therefore that there is no significant survival advantage for either of the described sequences, at least for the particular drug and radiation protocol investigated.

Some investigators have examined the results and toxicities of treatment regimens where radiotherapy and chemotherapy are given concurrently. Whereas Johnson et al. (1978) reported severe toxicity with radiation doses of about 30 Gy in 15 fractions, Greco et al. (1978), giving 30 Gy in 10 fractions, Cohen et al. (1980), giving 40 Gy in 15 fractions, and Perry et al. (1981) giving 50 Gy in 25 fractions did not consider toxicity to be a major problem, and achieved acceptable results. A prospective randomised study has been commenced in the United States by the Cancer and Leukaemia Group B to investigate further the feasibility of giving radiotherapy and chemotherapy simultaneously.

Local Radiotherapy in Conjunction with Wide-Field Irradiation. By analogy with the administration of chemotherapy to treat micro- or occult metastases in early disease, it might be argued that treating large fields with irradiation – such as total-body irradiation (TBI) or sequential half-body irradiation (HBI) – would serve the same purpose. Many of the trials in which this technique is being tested are in their early stages, and few have concentrated on the results in limited disease (the majority of studies having mixed groups of patients or only patients with extensive disease).

Dawes (1980) has studied 22 patients, giving TBI or two HBIs without any other treatment in doses of 1.5 Gy in two fractions or 15 Gy in ten fractions for TBI and 8 Gy for each HBI. Patients with localised disease had a median survival of 5 months (and those with extensive disease, $2^{1}/_{2}$ months). Urtasun et al. (1982) reported more favourable findings in a randomised study of 64 patients comparing chemotherapy and hemibody irradiation in patients who also had radiotherapy to the primary site in the chest. The upper HBI resulted in a single midplane dose of 8 Gy (lungs shielded to receive 6 Gy corrected for lung transmission) and the lower HBI was given 6 weeks later. Patients with early disease showed response rates of 94% and 89% with HBI and chemotherapy, respectively, and the estimated median survival times were 43 and 42 weeks, respectively. The authors suggest that sequential treatment with HBI and local radiation boost is an efficient method of tumour control in patients with early disease. For advanced disease the response rates with HBI and chemotherapy were 77% and 87%, respectively, with estimated median survivals of 15 and 44 weeks, respectively. Thus the same conclusions cannot be drawn for advanced disease as for limited disease. A further study by Urtasun et al. (1983) in 26 patients, with fractionated HBI (10 Gy in four fracitons) given to consolidate and maintain responses induced by chemotherapy, shows an overall median survival of 14 months for local and extensive disease. Failure in the primary site was seen in all relapsing patients, and there were appreciable failures in liver, brain, and upper abdomen. Because of this the authors have discontinued their use of lower HBI and have increased their dose to the upper body half to 20 Gy in eight fractions. Other randomised studies by Woods et al. (1981) and by Byhardt et al. (1979) suggest marginally shorter survival times in patients receiving HBI or TBI than in those receiving conventional chemotherapy.

However, no long-term survival data are available for these studies, and although there might be advantages for TBI and HBI over chemotherapy (provided equivalent results were obtained), it does not seem so far that large-field irradiation in limited small cell lung cancer is particularly promising.

Prophylactic Radiotherapy

The central nervous system (CNS) has a high predilection for metastatic disease from small cell lung cancer, and Bunn and Ihde (1981) have reported a 22% relapse rate in the brain. Their study was nonrandomised and retrospective, but is showed that the relapse rate was only 8% in these patients who had received prophylactic cranial irradiation (PCI). This study, however, showed no improvement in the median survival time — a finding contrary to that described by Rosen et al. (1981) in another retrospective study. This later analysis showed a statistically significant improvement in the 2-year survival rate (18%−20% following PCI; 5% without PCI). The question of whether PCI improved survival and reduced the incidence of cerebral metastases prompted several centres to perform randomised studies to examine this problem. Of seven studies (Jackson et al. 1977; Maurer et al. 1980; Hirsch et al. 1979; Beiler et al. 1979; Cox et al. 1978; Seydel et al. 1981; Aisner et al. 1982), two (Hirsch et al. 1979; Cox et al. 1978) showed no significant difference in the percentage relapse rate. The studies by Hirsch et al. (1979) and Seydel et al. (1981) were based on patients with limited disease only, whereas that of Aisner et al. (1982) was the only one to randomise only complete responders. The cumulative results in a total of 402 patients suggest a 20% incidence of brain metastases in the group not receiving PCI — figures almost identical to those of Rosen et al. (1981) and Bunn and Ihde (1981). However, there was no significant improvement in survival in any study, but a retrospective study of Rosenman and Choi (1982) reports an enhanced quality of life in patients given brain irradiation prophylactically rather than therapeutically. The question thus remains as to whether the brain should be treated prophylactically or only therapeutically when cerebral metastases are confirmed.
There is also some doubt as to the optimum timing of PCI and the dose. The above-mentioned studies quote a dose range of between 20 Gy and 40 Gy over 2−4 weeks, and PCI was given at any time between the first day of treatment and the twelfth week. The analysis by Rosen et al. (1981) indicated that patients who do not achieve a complete response have a very high likelihood of developing CNS metastases (up to 100% at 2−3 years) whether or not PCI is given. Aisner's randomised study (Aisner et al. 1982) examined the effect of PCI only in patients achieving a complete response, but the statistical significance of the survival figures for that study was not reported.
There is still no answer therefore as to whether PCI is of value in prolonging survival in complete responders, but it seems from the results of Rosen et al. (1981) that the stage of the disease and liver involvement are the most important CNS risk factors. Further questions also arise, such as the role of prophylactic craniospinal irradiation: Holoye et al. (1983) described a 14% intraspinal relapse rate in patients with limited and extensive disease treated with chemotherapy, PCI, and consolidation X-ray treatment to the chest. The answers to these questions can only be obtained if further information can be obtained by randomised prospective trials.
It is the consensus of the International Association for the Study of Lung Cancer Group (Bleehen et al. 1983) that, as these patients have a short survival (99% dying within 3 years), and as therapeutic brain irradiation provides good symptomatic relief, PCI should not be given to patients with extensive disease unless they achieve a complete response.

Radiotherapy in Extensive Disease

The value of radiotherapy in a palliative setting is well recognised in extensive small cell lung cancer. But what is its local role in conjunction with chemotherapy or in large-field radiotherapy? The results described by Urtasun et al. (1982) are worse for extensive disease than for limited disease (the estimated median survival after sequential HBI + intrathoracic radiation was 15 weeks, compared with 44 weeks for chemotherapy + intrathoracic irradiation). Dillman et al. (1983) have examined the effect of chemotherapy, followed by TBI, PCI, and non-cross-resistant chemotherapy (given in that order) in a nonrandomised study of 21 patients with extensive disease. They found that TBI does not appear to add substantially to the therapeutic benefit of combination chemotherapy alone in extensive small cell lung cancer. Thus it seems that locoregional radiotherapy has a limited role in extensive disease when it is not in a palliative context.
Is it necessary to give intrathoracic radiation, or PCI, in patients with extensive disease? The discussion above has indicated that PCI is not indicated in this context, and there are very few trials to assess the role of radiation to the primary site in addition to chemotherapy. Williams et al. (1977) showed that the addition of locoregional radiotherapy to chemotherapy plus PCI did not alter the meidan survival time or change the frequency of relapse.
There is no definitive answer to the question of the role of radiotherapy in extensive disease treated by chemotherapy. However, with improved ontrol of disseminated disease by newer drugs, it may be that radiation will have a role in the control of bulky intrathoracic primary disease.

Palliative Radiotherapy

Radiotherapy is of undoubted value in controlling the symptoms of bronchial obstruction, metastatic bone pain, cerebral metastases, spinal cord compression, and superior vena caval obstruction due to small cell lung cancer. The dose and fractionation depends on the site of the disease and on the preference of the radiotherapist.
Baglan and Marks (1981) have reported that 92% of patients receiving brain irradiation therapeutically had excellent palliation of symptoms. Crane et al. (1983) found retrospectively in 153 patients that 32% developed brain metastases. They believe that cranial radiotherapy reduces morbidity from brain metastases so that the majority of patients die from noncerebral metastatic disease. The dose range was 30−44 Gy according to disease status and the time that brain metastases developed, and they suggest that patients with brain metastases at diagnosis or presenting with brain metastases as the sole site of metastatic disease require the higher doses of radiation.
Chemotherapy is also of value in the palliation of widespread disease, but there is no hard evidence concerning the relative merits of the two treatment modalities for the palliation of local symptoms such as superior vena cava obstruction.

Conclusions

For limited disease, radiotherapy to locoregional disease may add to long-term control of disease at that site, although firm evidence in terms of survival is still awaited; for extensive disease this does not seem to be the case.

Prophylactic cranial irradition does not seem to be of value in patients with extensive disease, and even though it reduces the incidence of cerebral metastasis its value has not been defined when given to patients with limited disease achieving complete response.

The dose of radiation for locoregional disease is not defined, but is probably lower in conjuction with chemotherapy. The optimal timing of radiation within a chemotherapy regimen is not known.

Hemibody and total-body irradiation may well have palliative roles in extensive disease, but their value instead of chemotherapy in limited disease is still under study.

Chemotherapy now seems to be the major treatment modality for small cell lung cancer. The role of radiotherapy, which used to be the mainstay of treatment, is now being revised.

References

Aisner J, Whitacre M, Van Echo DA, et al. (1982) Combination chemotherapy for small cell carcinoma of the lung: continuous vs alternating non-cross-resistant combinations. Cancer Treat Rep 66: 221−230

Baglan RJ, Marks JE (1981) Comparison of symptomatic and prophylactic irradiation of brain metastases from oat cell carcinoma of the lung. Cancer 47: 41−45

Beiler DD, Kane RC, Bernath AM et al. (1979) Low dose elective brain irradiation in small cell carcinoma of the lung. Int J Radiat Oncol Biol Phys 5: 944−945

Bergsagel DE, Jenkin FDT, Pringle JD et al. (1972) Lung cancer: clinical trial of radiotherapy alone vs radiotherapy plus cyclophosphamide. Cancer 30: 621−627

Bleehen NM (1979) Role of radiation therapy and other modalities in the treatment of small cell carcinoma of the lung. In: Muggia F, Rozenweig M (eds) Lung cancer progress in therapeutic research. Raven, New York, pp 567−574

Bleehen NM, Bunn PA, Cox JD et al. (1983) Role of radiation therapy in small cell anaplastic carcinoma of the lung. Cancer Treat Rep 67: 11−19

Bunn PA, Ihde DC (1981) Small cell bronchogenic carcinoma: a review of therapeutic results. In: Livingston RB (ed) Lung cancer I. Martinius Nijhoff, The Hague, pp 169−208

Byhardt RW, Cox JD (1983) Is chest radiotherapy necessary in any or all patients with small cell carcinoma of the lung? Yes. Cancer Treat Rep 67: 209−215

Byhardt RW, Cox JD, Wilson JF et al. (1979) Total body irradiation vs chemotherapy as a systemic adjuvant for small cell carcinoma of the lung. Int J Radiat Oncol Biol Phys 5: 2043−2048

Choi CH, Carey RW (1976) Small cell anaplastic carcinoma of lung. Cancer 37: 2651−2657

Cohen MH (1983) Is thoracic radiation necessary for patients with limited-stage small cell lung cancer? No. Cancer Treat Rep 67: 217−221

Cohen MH, Lichter AS, Bunn PA et al. (1980) Chemotherapy-radiation therapy versus chemotherapy in limited stage small cell lung cancer. Proc Am Assoc Cancer Res ASCO 21: 448

Cox JD, Petrovich Z, Paig C et al. (1978) Prophylactic cranial irradiation in patients with inoperable carcinoma of the lung. Cancer 42: 1135−1140

Crane J, Lichter A, Ihde D, Carney D, Gold GL, Glatstein E, Minna J, Bunn P (1983) Therapeutic cranial radiotherapy for brain metastases in small cell lung cancer. Proc Am Assoc Cancer Res 24: 145

Daddi G, Di Giorgio A, Sammartino P, Naticchioni E, Biocaa P (1982) Small cell carcinoma of the lung: immediate and long-term results after surgical treatment. In: Abstracts of the III world converence on lung cancer, Tokyo, p 167

Dawes PJDK (1980) Results of a pilot study of wide field radiotherapy in the treatment of oat cell carcinoma of the bronchus. Clin Radiol 31: 723−727

Deeley TJ (1966) A clinical trial to compare two different tumour dose levels in the treatment of advanced carcinoma of the bronchus. Clin Radiol 17: 299–301

Dillman RO, Seagren SL, Beauregard JC, Teatle R (1983) Combination chemotherapy and total body irradiation in extensive small cell carcinoma of the lung. Proc Am Soc Clin Oncol 2: 191

Dombernowsky P, Hansen HH, Hansen M et al. (1980) Treatment of small cell anaplastic bronchogenic carcinoma. Results from 2 randomized trials (abstr). II world conference on lung cancer, Copenhagen. Excerpta Medica, Amsterdam, p 149

Einhorn LH, Bond WH, Hornback W et al. (1978) Long term results in combined modality treatment of small cell carcinoma of the lung. Semin Oncol 5: 309–313

Fox W, Scadding JG (1973) Medical Research Council comparative trial of surgery and radiotherapy for primary treatment of small-celled or oat-celled carcinoma of bronchus: ten-year follow-up. Lancet 2: 63–65

Fox RM, Woods RL, Brodie GN et al. (1980) A randomized study: small cell anaplastic lung cancer treated by combination chemotherapy and adjuvant radiotherapy. Int J Radiat Oncol Biol Phys 6: 1083–1085

Greco FA, Richardson RL, Shulman SF, Stroup S, Oldham RK (1978) Therapy of oat cell carcinoma of the lung: complete remissions, acceptable complications and improved survival. Br Med J 2: 10–11

Greco FA, Richardson RL, Snell JD et al. (1979) Small cell lung cancer: complete remission and improved survival. Am J Med 66: 625–630

Harper PG, Souhami RL, Spiro SG, Geddes DM, Tobias JS (1983) Chemotherapy with and without radiotherapy in small cell carcinoma of the lung. Proc Am Assoc Cancer Res 24: 151

Higgins GA, Shields TW, Matthews MJ (1982) The role of surgery in small cell carcinoma of the lung. In: Abstracts of the III world conference on lung cancer, Tokyo, p 165

Hirsch FR, Hansen HH, Paulson OB et al. (1979) Development of brain metastases in small cell anaplastic carcinoma of the lung. In: Kay J, Whithouse J (eds) CNS complications of malignant disease. Macmillan, New York, pp 175–184

Holoye PY, Samuels ML, Lanzottie VJ et al. (1977) Combination chemotherapy and radiation therapy for small cell carcinoma. JAMA 237: 1221–1224

Holoye PY, Samuels ML, Smith T et al. (1978) Chemoimmunotherapy of small cell bronchogenic carcinoma. Cancer 42: 34–40

Holoye P, Libnoch J, Byhardt R, Cox J (1983) Treatment of small cell bronchogenic carcinoma with high dose vincristine chemotherapy and consolidation radiation therapy. Proc Am Soc Clin Oncol 2: 186

Høst H (1973) Cyclophosphamide (NSC-26271) as adjuvant to radiotherapy in the treatment of unresectable bronchogenic carcinoma. Cancer Chemother Rep 4: 161–164

Jackson DV, Richards F, Cooper MR et al. (1977) Prophylactic cranial irradiation in small cell carcinoma of the lung. A randomized study. JAMA 237: 2730–2733

Johnson RE, Brereton HD, Kent CH (1978) Total therapy for small cell carcinoma of the lung. Ann Thorac Surg 25: 510–515

Karrer K, Denck H, Pridun N, Zwintz E (1982) Positive effect of combined treatment for small-cell bronchial carcinoma using surgery for cure and polychemotherapy. In: Abstracts of the III world conference on lung cancer, Tokyo, p 164

Krauss S, Perez C (1979) Treatment of localized undifferentiated small cell lung carcinoma (SCLC) with radiation therapy (RT) with or without combination chemotherapy (CT) with cyclophosphamide (C), adriamycin (A) and dimethyltriazenoimidazole carboxamide (DTIC). Proc Am Assoc Cancer Res ASCO 20: 316

Levison V (1980) Pre-operative radiotherapy and surgery in the treatment of oat cell carcinoma of the bronchus. Clin Radiol 31: 345–348

Levison V (1982) How should we treat early operable small cell carcinoma of the bronchus? In: Abstracts of the III world conference on lung cancer, Tokyo, p 166

Livingston RB, Moore TN, Heilbrun L et al. (1978) Samll cell carcinoma of the lung: combined chemotherapy and radiation. Ann Intern Med 88: 194–199

Matthiessen W (1978) Controlled clinical trial of radiotherapy alone, against radiotherapy plus chemotherapy in small-cell carcinoma of the lung: comparison of radiation damage (preliminary results). Scand J Respir Dis 59: 209–211

Maurer LH, Tulloh M, Weiss RB et al. (1980) A randomized combined modality trial in small cell carcinoma of the lung: comparison of combination chemotherapy-radiation therapy versus cyclophosphamide-radiation therapy, effects of maintenance chemotherapy and prophylactic whole brain irradiation. Cancer 45: 30–39

Medical Research Council Lung Cancer Working Party (1979) Radiotherapy alone or with chemotherapy in the treatment of small-cell carcinoma of the lung. Br J Cancer 40: 1–10

Medical Research Council Lung Cancer Working Party (1981) Radiotherapy alone or with chemotherapy in the treatment of small-cell carcinoma of the lung: the results at 36 months. Br J Cancer 44: 611–617

Medical Research Council Lung Cancer Working Party (1983) Cytotoxic chemotherapy before and after radiotherapy compared with radiotherapy followed by chemotherapy in the treatment of small-cell carcinoma of the bronchus: the results up to 36 months. Br J Cancer 48: 755–761

Miller AB, Fox W, Tall R (1969) Five-year follow-up of the Medical Research Council comparative trial of surgery and radiotherapy for the primary treatment of small-celled or oat-celled carcinoma of the bronchus. Lancet 2: 501–505

Mouritzen C, Fasting H, Jakobsen BM (1982) Five years survival after surgical treatment for small cell carcinoma of the lung. In: Abstracts of the III world conference on lung cancer, Tokyo, p 164

Nakagawa K, Matsubara T, Kinoshita I, Tsuchiya E (1982) Application of resection to small cell carcinoma. In: Abstracts of the III world conference on lung cancer, Toky, p 165

Ohta M, Hara N, Nakada T et al. (1982) Survival of patients with resected small cell carcinoma. In: Abstracts of the III world conference on lung cancer, Tokyo, p 166

Perez CA (1977) Radiation therapy in the management of carcinoma of the lung. Cancer 39: 901–916

Perez CA, Einhorn L, Oldham RK, Cohen H, Silberman H, Krauss S, Hornback N, Comas F, Greco FA, Birch R, Dandy M (1983) Preliminary report on a randomised trial of radiotherapy to the thorax in limited small cell carcinoma of the lung treated with multiagent chemotherapy. Proc Am Soc Clin Oncol 2: 190

Perry MC, Eaton WL, Comis RL, Spaulding MB, Carey RW, Maurer LH (1981) Simultaneous radiation therapy and chemotherapy in limited small cell cancer of the lung: a pilot study. Proc Am Soc Clin Oncol 22: 503

Petrovich Z, Mietlowski W, Ohanion M et al. (1977) Clinical report on the treatment of locally advanced lung cancer. Cancer 40: 72–77

Rosen S, Bunn PA, Lichter A et al. (1981) Prophylactic cranial irradiation (PCI) in small cell lung cancer (SCLC): benefit restricted to patients in complete response. Proc Am Assoc Cancer Res ASCO 22: 499

Rosenman J, Choi N (1982) Improved quality of life of patients with small-cell carcinoma of the lung by elective irradiation of the brain. Int J Radiat Oncol Biol Phys 8: 1041–1043

Roswit B, Patno ME, Rapp R, Veinbergs A, Feder B, Stuhlbarg J, Reid CB (1968) The survival of patients with inoperable lung cancer. A large-scale randomized study of radiation therapy versus placebo. Am J Roentgenol 90: 688–697

Rubin P, Perez CA, Keeler B (1976) The logical basis of radiation treatment policies in the multidisciplinary approach to lung cancer. In: Israël L, Chakinian AP (eds) Lung cancer. Academic Press, New York, pp 159–197

Salazar OM, Creech RH (1980) "The state of the art" towards defining the role of radiation therapy in the management of small cell bronchogenic carcinoma. Int J Radiat Oncol Biol Phys 6: 1103–1117

Seydel HG, Creech R, Pagano M et al. (1981) Combined modality treatment of small cell undifferentiated carcinoma of the lung. A cooperative study of the RTOG and the ECOG. Int J Radiat Oncol Biol Phys (Suppl) 7: 41

Souhami RL, Harper PG, Linch D, Trask C, Gladstone AH, Tobias JS, Spiro SG, Geddes DM,

Richards JDM (1983) High dose cyclophosphamide with autologous marrow transplantation for small cell carcinoma of the bronchus. Cancer Chemother Pharmacol 10: 205–207

Stevens E, Einhorn L, Rohn R (1979) Treatment of limited small cell lung cancer. Proc Am Assoc Cancer Res ASCO 20: 435

Tucker RD, Sealy R, Van Wyk C et al. (1973) A clinical trial of cyclophosphamide (NSC-26271) and radiation therapy for oat cell carcinoma of the lung. Cancer Chemother Rep 4: 159–161

Urtasun R, Belch A, McKinnon S et al. (1982) Small-cell lung cancer: initial treatment with sequential hemi-body irradiation vs 3 drug systemic chemotherapy. Br J Cancer 46: 228–235

Urtasun R, Belch A, Bodmer D (1983) Toxicity and patterns of failure of hemibody irradiation as a consolidating agent for patients with small cell lung cancer. Proc Am Soc Clin Oncol 2: 187

Williams C, Alexander M, Glatstein EJ et al. (1977) Role of radiation therapy in combination with chemotherapy in extensive oat cell cancer of the lung: a randomized study. Cancer Treat Rep 61: 1427–1431

Wolf J, Patno ME, Roswit B, D'Esopo N (1966) Controlled study of survival of patients with clinically inoperable lung cancer treated with radiation therapy. Am J Med 40: 360–267

Woods RL, Tattersall MHN, Fox RM (1981) Hemi-body irradiation (HBI) in "poor prognosis" small cell lung cancer (SCLC). Proc Am Assoc Cancer Res ASCO 22: 502

Chemotherapeutic Results in Small Cell Lung Cancer

N. Niederle and J. Schütte

Innere Universitätsklinik und Poliklinik (Tumorforschung), Westdeutsches Tumorzentrum, Hufelandstrasse 55, 4300 Essen, Federal Republic of Germany

Introduction

Small cell lung cancer (SCLC) is a distinct and well-reproducible histopathological entity (Hirsch et al. 1982). Besides morphological criteria, SCLC is distinguished from other cell types by ultrastructural, endocrine, cytogenetic, cell-kinetic, and clinical findings (Gropp et al. 1980; Hattori et al. 1972; Whang-Peng et al. 1982). The high labeling index (Muggia et al. 1974; Pettengill et al. 1980) and short doubling time (Straus 1974; Weiss et al. 1970) are manifested clinically by fast growth, rapid onset of signs and symptoms, and a tendency toward early and widespread dissemination (Hansen et al. 1978; Muggia and Chevru 1974). Accordingly, the median survival time from diagnosis without treatment is 2.8 months, with only 4.2% of patients still alive at 1 year (Hyde et al. 1965; Zelen 1973).
Since SCLC must be considered to be disseminated at the time of diagnosis in the majority of patients, local forms of treatment such as surgery or radiotherapy alone have proven to be of little benefit. In most patients thus treated, disease recurred within only a few months (Mountain 1974; Wolf et al. 1966). Only when effective cytostatic drugs became available did median and 1-year survival rates improve significantly (Table 1).

Activity of Single Drugs

A statistically significant increase in survival time for SCLC patients treated with cyclophosphamide as compared to placebo was first reported in 1969 (Green et al.). Over the following years, SCLC was established as the most sensitive of all histologic types of lung cancer, though in many of these studies only a few patients were included. In addition, the information about the patients included is sometimes insufficient to form reliable conclusions concerning the true activity of the compound. Moreover, as most of the patients were extensively pretreated with other cytostatic agents, newer drugs (e.g., etoposide, cisplatin, vindesine) have been tested at a great disadvantage.
In spite of this, a growing number of drugs with clinical activity has been identified (Table 2). According to a review by Broder and co-workers (1977), which has since been updated (Niederle 1983), the response frequency to CCNU, 5-fluorouracil, procarbazine, and bleomycin has been found to be rather unsatisfactory, whereas the more active compounds are adriamycin, cisplatin (all patients were pretreated with other active drugs) cyclophosphamide, ifosfamide, hexamethylmelamine, etoposide, methotrexate (no significant differences between high and low dose), vincristine, and vindesine. By utilizing the latter drugs, objective response rates in the range of 30%−50% can be achieved. But these responses have been rarely (0%−8%) complete and generally transient, with median

Recent Results in Cancer Research. Vol. 97
© Springer-Verlag Berlin · Heidelberg 1985

Table 1. Median and 1-year survival in patients with SCLC following radiotherapy (RT) or radiotherapy plus chemotherapy (RT + CT)

Treatment	Patients n	Median survival		1-year survival (%)	Reference
		RT (months)	RT+CT (months)		
4,000–5,000 rads	14	5	–	–	Bergsagel et al.
4,000–5,000 rads + Cy	27	–	10	–	(1972)
4,000–5,000 rads	27	–	–	19	Carr et al. (1972)
4,000–5,000 rads + F	31	–	–	23	
4,000 rads	36	5.5	–	25	Høst (1973)
4,000 rads + Cy	39	–	8.5		
5,000–6,000 rads	34	5	–	28	Petrovitch et al. (1977)
5,000–6,000 rads + CCNU, Hy	35	–	9.5		
4,500 rads + CT by progress	21	5.5	–	–	Krauss and Perez
4,500 rads + A, Cy, D	32	–	11	–	(1980)
3,000 rads	121	6	–	18	Medical Research
3,000 rads + Cy, CCNU, M	115	–	10	34	Council (1979)

Cy, cyclophosphamide; *F*, 5-fluorouracil; *CCNU*, lomustine; *A*, adriamycin; *M*, methotrexate; *D*, dacarbazine; *Hy*, Hydroxyurea

Table 2. Activity of single drugs in SCLC. (CR, complete remission)

	Patients (n)	CR (%)	Response (%)
Adriamycin	53	4	30
CCNU	76	4	14
Cisplatin[a]	110	1	16
Cyclophosphamide	389	–	39
Etoposide	288	5	37
Hexamethylmelamine	69	7	30
Ifosfamide	28	–	71
Methotrexate	73	–	30
Procarbazine	43	–	21
Vincristine	43	7	42
Vindesine	47	4	32

[a] All patients pretreated

survival times below 20 weeks. The findings could be significantly improved by combinations of single antineoplastic agents.

Combination Chemotherapy

Definite activity in SCLC has been shown for a large number of chemotherapeutic regimens combining two or more drugs having different mechanisms of action and no

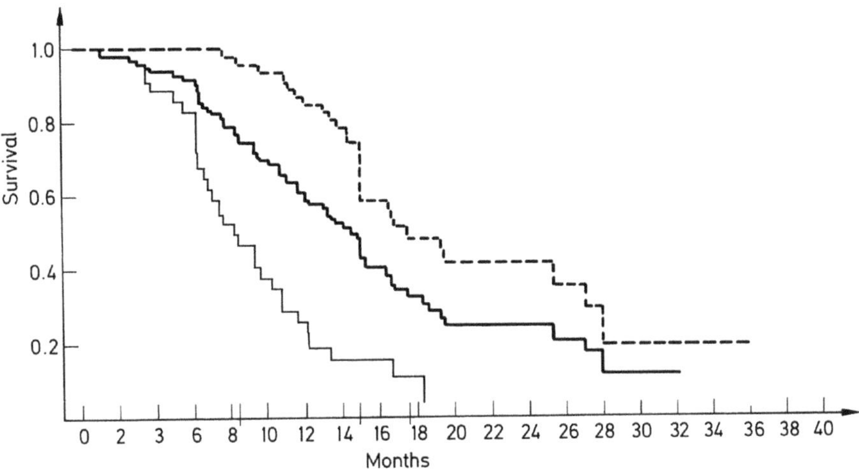

Fig. 1. Actual survival times for 100 patients with SCLC; ———, all patients; ------, complete responders; ——— less than complete responders (Niederle et al. 1982b)

major overlapping toxicity. The superiority of combination chemotherapy over single agents is well established (Lowenbraun et al. 1979; Maurer et al. 1980; Vincent et al. 1981). In addition, simultaneous application of drugs seems to be superior to the sequential use of each of the compounds (Alberto et al. 1976).

In most studies the stage of disease – either limited or extensive – and the performance status have been reported to be the major factors influencing prognosis. Further prognostic factors are number and site of metastases, pretreatment serum lactate dehydrogenase (LDH) values and the extent of previous chemotherapeutic exposure. Moreover, median survival times are markedly better for patients achieving complete remissions than for partial responders (Fig. 1), both being calculated using standard criteria (Miller et al. 1981). On the other hand, most studies indicate that subtyping of SCLC using the WHO criteria has no clinical importance (Burdon et al. 1979, Matthews 1983; Strauchen et al. 1983). Nevertheless, patients with tumors containing large cell components should be evaluated separately, since these mixed tumors seem to respond less favorably to therapy than 'pure' SCLC (Hirsch et al. 1983; Radice et al. 1982). Based on the largest series reported, the results of intensive combination chemotherapy using three or more drugs can be summarized as follows.

Extensive Disease

In extensive disease patients, complete remissions usually lasting 4–7 months can be induced in 20%–50% of patients (Table 3). The median survival time averages 7–11 months. One-year survival rates are in the range of 20%–40%, while less than 10% of patients survive more than 2 years. However, long-term disease-free survival can be obtained in a very low fraction of patients (Matthews et al. 1980; Peschel et al. 1981; Smith et al. 1981). In most trials, treatment is continued for at least 6–12 months or even until relapse, with a crossover to other mutually non-cross-resistant regimens. Additional irradiation to the primary disease sites or the cranium has been infrequently included, since radiotherapy has little effect on survival in patients with extensive disease.

Table 3. Results of combination chemotherapy for extensive disease patients

Treatment			Patients (n)	Remissions (%)	CR (%)	Median survival (months)	Survival at years (%)			Reference
Chemotherapy	X-ray (Gy)						1	2	3	
	Mediastinum and primary tumor sites	Cranium								
A,Cy,V(M)	30	30	250	56	14	6	16	–	–	Livingston et al. (1978)
A,Cy,V	40	–	22	59	31	7	35	0	–	Saugier et al. (1978)
B,Cy,MeCc,V	–	–	26	65	–	7	20	0	–	Van Houtte et al. (1979)
Cy,M,Cc(2x)// A,V,Pr(2x)±E,I	–	–	42	91	36	9.3	~39	20	–	Cohen et al. (1979)
A,Cy,V (−PD)	–	–	24	–	50	10	30	0	–	Seeber et al. (1980)
A,Cy,V (9x)	$<{}^{40}_{\varnothing}$	– –	25 27	74	20	9.5 9	– –	– –	– –	Fox et al. (1980)
A,Cy,V	–	–	29	–	59	8.5	40	7	–	Catane et al. (1981)
Cy,M,V//A,E (−PD)	–	–	66	72	30	8	–	5	–	Aroney et al. (1982)
A,Cy,V (4x) →Cc,M,E (4x)	–	–	36	83	33	9.3	38	0	–	Niederle et al. (1982b)
$<$Cy,E,A (−2y) Cy,E,A//Cc,M,V,Pr (−2y)	–	CR$<{}^{30}_{\varnothing}$	32 33	}86	40	11 8	27	0	–	Aisner et al. (1982)

H,Cy,A,V (2x) → Cy,M,E//A,M,E (6x)	40	20	34	61	35	10	–	–	–	Young et al. (1982)
M,Cy,A,V//E,V,H	–	30–39	40	70	25	9	–	8 (1.5y)	–	Hande et al. (1982)
<Cc,Cy,V,M→E,A (at PD)	–	–	66	68	17	8	–	~ 8	–	Østerlind et al.
Cc,Cy,V,M//E,A	–	–	64	72	23	9	–	~ 8	6 (4y)	(1983)
Cy,A,V	–	} 30 at CR	46	52	7	7	–	–	–	Issell et al.
<Cy,V (14x)	–		36	49	8	7	–	–	–	(1983)
Cy,E,V	–		48	60	18	8	–	–	–	

CR, complete remission; *PD*, progressive disease; *A*, adriamycin; *B*, bleomycin; *Bc*, BCNU; *Cc*, CCNU; *Cy*, cyclophosphamide; *D*, dacarbazine; *E*, etoposide; *F*, 5-fluorouracil; *H*, hexamethylmelamine; *Hy*, hydroxyurea; *I*, ifosfamide; *M*, methotrexate; *MeCc*, methyl-CCNU; *Mt*, mitomycin; *P*, cisplatin; *Pr*, procarbazine; *V*, vincristine; *VLB*, vinblastine; *VDS*, vindesine; //, alternating chemotherapy; /, sequential chemotherapy; −2y, therapy up to 2 years; →, followed by; <, ≤, prospective randomized

Table 4. Combination chemotherapy for limited disease patients

Treatment			Patients (n)	Remissions (%)
Chemotherapy	X-ray (Gy)			
	Mediastinum, primary tumor sites	Cranium		
A,Cy,V (4–5x)	30	20–30	11	–
A,Cy,V (M) (>9x)	45	30	108	75
A,Cy,V	40	–	16	75
A,Cy,V (6x),E,H (3x)	30	30	32	100
Cy,M,Cc,V	$<{}^{40}_{\ 0}$	– –	65 69	87 92
B,Cy,M,MeCC (9x)	45	–	58	83
A,Cy(Cy,M,V) (−24 m)	$<{}^{30}_{40-45}$	30	10	100
A,Cy,V(M) (>9x)	$<{}^{35}_{\ 0}$	35	14 18	100 75
A(M)Cy,E (−2y)	–	–	11	100
$<$ A,Cy,E,V, A,Cy,E,V,P	40	–	30 31	– –
A,Cy,V,(M) (>10x)	$<{}^{40}_{\ 0}$	– –	36 37	87
Cy,M,Cc (−18 m)	40	– 40 (+ retroperitoneal lymph nodes)	55 54	68 64
A,Cy,V	40	40	24	83
A,Cy,V(M) (>9x)	30	30	26	–
A,Cy,V (4x)	−45	–	42	–
$<{}^{0}$ Cy,M,Cc (>10x)	30	– –	121 115	– –
Cy,M,F,V A,Cy,V $\pm$ BCG	45	30	97 112	71 (after *)
Cy,Cc,M (−PD)	20 (UHBi 6Gy)	–	8	75
A,Cy,V (4x)	30–36	30	64	91
M-$\cancel{t}$-Cy (3 m)	30	–	55	78
Cy,M,Pr,V (−PD)	40–50	–	21	95
Cy,Cc (>2y)	36–45	30	217	80

UHBi, Upper hemibody irradiation; $\cancel{t}$, radiotherapy; *, remission duration

CR (%)	CR Duration (months)	Median survival (months)	Survival at years			Reference
			1	2	3	
			(%)			
100	10+	12+	80	–	–	Johnson et al. (1976)
41	11	12	50	25	–	Livingston et al. (1978)
45	–	20	70	55	–	Saugier et al. (1978)
90	11	16+	70	50	–	Greco et al. (1979)
–	6	11	–	–	–	Hansen et al. (1979)
–	6.5	14	–	–	–	
45	–	11.4	50	12	–	van Houtte et al. (1979)
80	–	$18.5<\genfrac{}{}{0pt}{}{16}{24}$	–	–	–	McMahon et al. (1979)
71	9.5	13	>30	–	–	Stevens et al. (1979)
50	8.5	11.5	>20	–	–	
73	11.5	14	–	–	–	Aisner and Wiernik (1980)
43	–	17	–	–	–	Eagan et al. (1980)
39	–	17	–	–	–	
21	–	16	69	–	–	Fox et al. (1980)
		14.5	64	–	–	
–	6	11.5	40	–	–	Hansen et al. (1980)
–	5.5	10.5	30	–	–	
58	–	–	71	–	–	Mercke et al. (1980)
77	9.5	21	60	27	–	Seeber et al. (1980)
81	–	13	60	37	25	Catane et al. (1981)
–	–	6	18	5	3	Medical Research Council (1981)
–		10	34	9	4	
7	–	10	–	–	–	McCracken et al. (1982)
12 (before ¼)	–	10	–	–	–	
37	(8)*	12	–	–	–	Mason et al. (1982)
72	9	15.8	68	33	11	Niederle et al. (1982b)
53	4	12	51	23	–	Thatcher et al. (1982)
57	–	13	59	20	–	Kessinger et al. (1983)
48	–	10	–	–	–	Seydel et al. (1983)

See Table 3 for definitions of other abbreviations

Table 5. Results of sequential (/) and alternating (//) chemotherapy in limited disease

Treatment			Patients (n)	Remissions (%)
Chemotherapy	X-ray (Gy)			
	Media-stinum, primary tumor sites	Cranium		
Cy,M,Cc (2x)//A,V,Pr (2×)±E,I	–	–	19	100
P,E (2×)//A,Cy,V (4×) (−18 m)	–	30	21	100
A,Cy,V//Bc,M,Pr (−2 y)	30	–	25	–
Cy,A,E (2×)//Bc,V,M,Pr (2×)	$<\,^{0}_{35}$	21	20	78 (LD+ED)
Cy,A,V/P,E (4×) $\overset{mc^{*}}{\Rightarrow}$Cc,M,Pr/Cy,A,V/E,P	50	40	24	100
E,V,A,/Cy,Cc,M	30	–	30	–
$<\genfrac{}{}{0pt}{}{Cy,E,A\ (-2\ y)}{Cy,E,A//Cc,M,Pr\ (-2\ y)}$	–	CR$<\,^{30}_{0}$	21 / 23	90
Cy,M,V//A,E (−PD)	–	–	32	81
Mt,M,E (3×)//A,Cy,V (3×)	30	30	34	93
A,P,E (2×)//Cc,Cy,V,M (2×)	35	30	15	87
H,Cy,A,V (2×) → Cy,M,E//A,M,E (6×)	40	20	22	82
$<\genfrac{}{}{0pt}{}{E,P\ (3\times)/A,Cy,V\ (3\times)}{A,Cy,V\ (3\times)/E,P\ (3\times)}$	CR $<\,^{40}_{0}$	30	27	92

See Tables 3 and 4 for definitions of other abbreviations
* Maintenance chemotherapy;   ** responders

Limited Disease

In limited disease patients (tumor confined to one hemithorax with or without mediastinal adenopathy and ipsilateral supraclavicular lymph node involvement), the treatment results have been more encouraging (Tables 4 and 5). With suitable combined drug regimens plus/minus radiotherapy, objective responses can be obtained in 70%−95% of previously untreated patients, including 40%−80% complete remissions. In most studies, median survival times are in the range of 10−15 months, increasing to approximately 20 months in complete responders. The percentage of patients living at 1 year is 40%−70%, and 5%−10% of all patients do achieve long-term remission of several years' duration, if not cure.

Sequence of Therapy. Since the various schedules of simultaneous combination chemotherapy (Table 4) have failed to show a significant increase in long-term survival

CR (%)	Median survival (months)		Survival at years			Reference
	all patients	CR patients	1	2	3	
			(%)			
74	14	–	55	12	–	Cohen et al. (1979)
52	16	–	60	40	–	Sierocki et al. (1979)
–	11.5	–	–	–	–	Sierocki et al. (1980)
40<55 (↯) / 22 (0 ↯)	16	25	70	30	–	Abeloff et al. (1981)
83	–	20**	73	24	–	Shank et al. (1981)
53	–	16.5	–	–	–	Vincent et al. (1981)
64	16 12	18	68	29	22	Aisner et al. (1982)
31	12+	13+	–	5	–	Aroney et al. (1982)
67	14	–	–	22	–	Murray et al. (1982)
73	12+	–	–	–	–	Vermorken et al. (1982)
68	14	15	–	–	–	Young et al. (1982)
63	12	13+(±↯)	40	10	–	Schütte et al. (1984)

during the last years, the concept of sequentially alternating chemotherapy (Goldie et al. 1982) has gained increased interest. Several groups have tried to prevent drug resistance by the early and predetermined application of an alternative drug schedule after an initial response to the first combination (Table 5). Within such sequences, each of the second regimens was able to improve the remission rates obtained with the first combination. However, sequentially alternating chemotherapy has so far not shown a significant increase in remission rates, remission duration, and median survival. Further follow-up and detailed analyses will reveal whether there is any positive impact on long-term survival.

Duration of Therapy. The more intensive and protracted a given chemotherapy, the higher the toxicity, and the worse the quality of life. This situation could be improved if, following a short yet intensive course of therapy, patients in complete remission were subjected to regular check-up only (Johnson et al. 1976; Niederle et al. 1982; Shank et al. 1981; Thatcher et al. 1982). So far, data on remission, duration of response, and survival are

Table 6. Chemotherapy and surgery

Treatment			Patients (n)	Remissions (n)
Chemotherapy	X-ray (Gy)			
	Media-stinum primary tumor sites	Cranium		
Stage I: OP→A,Cy,V,(E,Cc)	−	−	17 LD	−
Stage II: A,Cy,V,(E,Cc)→OP	−	−		−
OP$<$ Cc+Hy (−12 m)	−	−	11 LD	−
OP$<$ 0	−	−	18 LD	−
Cy	−	−	6 LD	−
OP≪Cy,M	−	−	, 9 LD	−
0	−	−	3 LD	−
OP→E,Cy,A,V,	50	30	8 LD	−
E,Cy,A,V,→OP			7 LD, 3 ED	−
A,Cy,V (2x)→OP→A,Cy,V	−	−	8 LD	8/8
A,Cy,V (3−6x)→OP→ A,Cy,V,(E,P) (3x)→⌀	+	+	5 LD	−
Cy,F,M,VLB OP≪Cy,Cy,M//A,Cy,V//I,E	30	2	26 Stage	−
0			14 I+II	−

OP, operation; *LD*, limited disease; *ED*, extensive disease; *NED*, no evidence of disease
See Tables 3 and 4 for definitions of other abbreviations

similar to results gained by continuous therapy. Moreover, this concept seems to be supported by prospective randomized studies concerning maintenance therapy: In patients achieving a complete remission, treatment beyond remission using induction regimens or alternative combinations had little if any impact on remission duration and survival (Bakker et al. 1982; Cullen 1982; Niederle et al. 1982b; Woods and Levi 1982).
The high rate of local relapse in SCLC implies, however, that many patients thought to be in complete remission on the basis of clinical presentation and radiographic findings in fact have residual disease, as has been demonstrated by serial fiberoptic bronchoscopies (Ihde et al. 1978; Nakhosteen and Niederle 1983; Sörenson 1983). Therefore, at present, short-term therapy can only be recommended to those patients for whom complete remissions suspected on the basis of clinical, radiological, and tumor-marker data are confirmed by more effective evaluations (e.g., bronchoscopy or even mediastinoscopy). In these patients, only four to six courses of chemotherapy should be administered, thus enabling them to enjoy a treatment-free period.

Radiotherapy. Though SCLC is known to be one of the most radiosensitive malignancies, the significance of radiotherapy as an adjunct to chemotherapy for limited disease still remains to be defined. Several studies have addressed the question of whether radiation

CR (n)	Median survival (months)		Survival at years				Reference
	all patients	CR patients	1	2	3	5	
			(%)				
–	20+	–	–	(13/17 NED)	–		Comis et al. (1982)
–		–	–	(11/17 NED)	–		
–	–	–	–	–	–	81	Shields et al. (1982a)
–	–	–	–	–	–	38	
–	7.7	–	–	–	–	16.7	
–	17.8	–	–	–	–	22.2	Shields et al. (1982b)
–	8.8	–	–	–	–	–	
–	27+	–	50 (NED)	–	–		Valdivieso et al. (1982)
–	14.5+	–	–	–	–	–	
7/8	15+	–	–	–	3/8	–	Gatzemeier (1983)
–	12–42+	–	(4/5 NED)	–	–		Ginsberg et al. (1983)
–	–	–	~65	33	33	33	Karrer et al. (1983)
–	–	–	~35	8	8	8 ·	

therapy to the chest encompassing the primary tumor and hilar and mediastinal lymph nodes can consolidate the cytostatic effects, thereby lowering the risk of recurrence and prolonging survival (Byhardt and Cox 1983; Cohen 1983), as demonstrated by McMahon and co-workers (1979). So far, available data derived from prospective randomized trials indicate a decrease of local recurrence following additional radiotherapy, while no major influence on median survival has been demonstrated, with the possible exception of partial responders following chemotherapy. Moreover, combined modality therapy may slightly increase the possibility of long-term survival (Abeloff et al. 1981; Fox et al. 1980; Hansen et al. 1979; Souhami et al. 1982). However, because of the increase in acute and chronic treatment-related morbidity, further follow-up revealing the rate of long-term survivors is required in order to form definite conclusions.

Central nervous system (CNS) involvement is a frequent complication of SCLC. None of the studies comparing prophylactic CNS irradiation (dose range 20–40 Gy) and no irradiation could detect significant differences in median survival, and the impact on long-term survival is still uncertain (Aisner et al. 1982; Aroney et al. 1983; Hansen et al. 1980; Jackson et al. 1977; Maurer et al. 1980). But in patients receiving prophylactic CNS irradiation, the time to brain relapse was significantly longer, and the incidence of CNS relapse was reduced from a median of 23% to 5% (Baglan and Marks 1981). Because of this well-established prophylactic activity, the improved quality of life (Rosenman and Choi 1982), and the minimal side effects (Catane et al. 1981), CNS irradiation might be applied – but only in patients achieving complete remission. They are the only potential long-term survivors, and the incidence of CNS relapse increases with the improvement of survival time (Burgess et al. 1979; Komaki et al. 1981; Rosen et al. 1983). In summary, the exact definition of the additional radiotherapeutic value remains to be assessed in trials answering basic questions concerning optimal dose-time relationships of radiotherapy when combined with effective chemotherapy.

Surgery. A further approach to the treatment of limited disease SCLC patients is the integration of surgery in the treatment plan. Total resection can be followed by adjuvant chemotherapy (Higgins 1972; Karrer et al. 1983; Shields et al. 1982a, b), or as an alternative, chemotherapy can be followed by thoracotomy to extirpate all visible disease, particularly in patients with greater bulk of disease (Comis et al. 1982; Gatzemeier et al. 1983; Ginsberg et al. 1983). To date, only preliminary results are available (Table 6). These data seem to warrant further investigation.

Relapse Therapy

In SCLC, remissions are difficult to maintain, and long-term survival still remains poor, as most patients relapse after 1–2 years (Aisner et al. 1982; Aroney et al. 1982; Niederle et al. 1982b; Young et al. 1982). Relapse after remission – as well as primary resistance to induction chemotherapy – implies rapid disease progress. Second or third remissions can only be induced by using non-cross-resistant cytostatic drugs and regimens. The results of different trials are listed in Table 7. Remission rates and median survival times are generally poor, probably due to extensive cytostatic pretreatment. Moreover, low responses in previously irradiated areas, at least in some studies (Niederle et al. 1984), illustrate that adjuvant radiotherapy after chemotherapeutically induced remissions might considerably compromise all second-line chemotherapeutic endeavors. In addition, in light of these results, the question of whether consolidating irradiation following chemotherapy provides any benefit in SCLC has to be reconsidered.

Conclusions

During the past decade, SCLC has been proven to be highly sensitive to various forms of nonsurgical therapy. The best results were obtained by applying more than two active antineoplastic agents simultaneously. So far, sequentially alternating non-cross-resistant combination chemotherapy, maintenance therapy, and adjuvant radiotherapy to primary disease sites or the brain have shown no major improvement in remission rates and median survival. It is still uncertain whether any positive impact on long-term survival can be substantiated through the use of these combined modalities of treatment.

As most responders relapse after 1–2 years, the elimination of resistant cell lines after initial response remains the main problem. New promising approaches might be:
– Development of new non-cross-resistant drugs
– Further intensification of chemotherapy employing ultrahigh doses of cytostatic drugs in combination with autologous bone marrow transplantation (see the chapter by Harper and Souhami, this volume)

Table 7. Second-line treatment after failure of prior chemotherapy

Treatment schedule	Patients (n)	Prior chemotherapy	Prior mediastinal irradiation (n)	Remissions (%)	CR (%)	Response duration (months)	Reference
E,P	20	A,Cy,V	–	50	10	4+	Osoba et al. (1981)
A	16	Cy,Cc	–	14	–	–	Ettinger and Lagakos (1982)
A,V	37	Cy,Cc	–	27	–	–	
A,E	56	Cy,Cc	–	30	–	–	
E (3x)	26	Cy,M,Cc	–	4	–	–	Harper et al. (1982)
Cc,M,E	23	A,Cy,V	11	26	9	7	Niederle et al. (1982a)
Cc,M,E,P	29	Cy,A,E	–	31	17	–	Poplin et al. (1982)
Cy,M,Cc, V,A,Pr	5	V,A,Cy,Cc,Pr, Cy,A,V,E	3	(4/5)	(2/5)	10	Batist et al. (1983)
VDS,P	32	A,Cy,V (E,Cc,M)	25	19	6	3	Niederle et al. (1984)
E,A	15	?	16	30	–	5	Roszkowski et al. (1983)
E,A,Cy	13	?	16	38	15	5	
E,A,P	12	?	16	41	16	5	

See Table 3 for definitions of abbreviations

- The combination of chemotherapy and surgery
- Other modalities of treatment such as late intensification chemotherapy, hyperthermia, anticoagulants, and maximal supportive care.

Moreover, in vitro models of SCLC should be used for diagnosis, drug sensitivity studies, and the identification of factors regulating tumor growth (Carney et al. 1983). The development of monoclonal antibodies (Ball et al. 1984; Cuttitta et al. 1981) with specificity for SCLC might offer further potential benefit in the clinical management of patients with SCLC.

References

Abeloff MD, Ettinger DS, Order SE, Khouri N, Mellits ED, Dorschel NT, Baumgardner B (1981) Intensive induction chemotherapy in 54 patients with small cell carcinoma of the lung. Cancer Treat Rep 65: 639−646

Aisner J, Wiernik PH (1980) Chemotherapy versus chemoimmunotherapy for small-cell undifferentiated carcinoma of the lung. Cancer 46: 2543−2549

Aisner J, Whitacre M, Van Echo DA, Wiernik PH (1982) Combination chemotherapy for small cell carcinoma of the lung: continuous versus alternating non-cross-resistant combinations. Cancer Treat Rep 66: 221−230

Alberto P, Brunner KW, Martz G, Obrecht JP, Sonntag RW (1976) Treatment of bronchogenic carcinoma with simultaneous or sequential combination chemotherapy, including methotrexate, cyclophosphamide, procarbazine and vincristine. Cancer 38: 2208−2216

Aroney RS, Bell DR, Chan WK, Dalley DN, Levi JA (1982) Alternating non-cross-resistant combination chemotherapy for small cell anaplastic carcinoma of the lung. Cancer 49: 2449−2454

Aroney RS, Aisner J, Wesley MN, Whitacre MY, Van Echo DA, Slawson RG, Wiernik PH (1983) Value of prophylactic cranial irradiation given at complete remission in small cell lung carcinoma. Cancer Treat Rep 67: 675−682

Baglan RJ, Marks JE (1981) Comparison of symptomatic and prophylactic irradiation of brain metastases from oat cell carcinoma of the lung. Cancer 47: 41−45

Bakker W, Noordijk EM, Van Oosterom AT (1982) Treatment of small cell lung cancer with combination chemotherapy and radiotherapy: no need for maintenance therapy following the induction of a complete response. Third World Conference on Lung Cancer, Tokyo (Abstr 206)

Ball ED, Graziano RF, Pettengill OS, Sorenson GD, Fanger NW (1984) Monoclonal antibodies reactive with small cell carcinoma of the lung. JNCI 72: 593−598

Batist G, Ihde DC, Zabell A, Lichter AS, Veach SR, Cohen MH, Carney DN, Bunn PA (1983) Small-cell carcinoma of the lung: reinduction therapy after late relapse. Ann Intern Med 98: 472−474

Bergsagel DE, Jenkin RDT, Pringle JF, White DM, Fetterly JCM, Klaassen DJ, McDermot RSR (1972) Lung cancer. Clinical trial of radiotherapy alone vs. radiotherapy plus cyclophosphamide. Cancer 30: 621−627

Broder LE, Cohen MH, Selawry OS (1977) Treatment of bronchogenic carcinoma. II. Small cell. Cancer Treat Rev 4: 219−260

Burdon JGW, Sinclair RA, Henderson MM (1979) Small cell carcinoma of the lung. Prognosis in relation to histologic subtype. Chest 76: 302−304

Burgess RE, Burgess VF, Dibella NJ (1979) Brain metastases in small cell carcinoma of the lung. JAMA 242: 2084−2086

Byhardt RW, Cox JD (1983) Is chest radiotherapy necessary in any or all patients with small cell carcinoma of the lung? Yes. Cancer Treat Rep 67: 209−215

Carney DN, Broder L, Edelstein M, Gazdar AF, Hansen M, Havemann K, Matthews MJ, Sorenson GD, Vindeløv L (1983) Experimental studies of the biology of human small cell lung cancer. Cancer Treat Rep 67: 27–35

Carr DT, Childs DS Jr, Lee RE (1972) Radiotherapy plus 5-FU compared to radiotherapy alone for inoperable and unresectable bronchogenic carcinoma. Cancer 29: 375–380

Catane R, Lichter A, Lee YJ, Brereton HD, Schwade JD, Glatstein E (1981) Small cell lung cancer: analysis of treatment factors contributing to prolonged survival. Cancer 48: 1936–1943

Cohen MH (1983) Is thoracic radiation therapy necessary for patients with limited-stage small cell lung cancer? No. Cancer Treat Rep 67: 217–221

Cohen MH, Ihde DC, Bunn PA Jr, Fossieck BE Jr, Matthews MJ, Shackney SE, Johnston-Early A, Makuch R, Minna JD (1979) Cyclic alternating combination chemotherapy for small cell bronchogenic carcinoma. Cancer Treat Rep 63: 163–170

Comis R, Meyer J, Ginsberg S, Issell B, Gullo J, DiFino S, Tinsley R, Poiesz B, Rudolph A (1982) The current results of chemotherapy (CTH) plus adjuvant surgery (AS) in limited small cell anaplastic lung cancer (SCALC). Proc Am Soc Clin Oncol 1: 147 (Abstr C-571)

Cullen MH (1982) Maintenance chemotherapy for small cell lung cancer: is it necessary? Third World Conference on Lung Cancer, Tokyo (Abstr 241)

Cuttitta F, Rosen S, Gazdar AF, Minna JD (1981) Monoclonal antibodies that demonstrate specificity for several types of human lung cancer. Proc Natl Acad Sci USA 78: 4591–4595

Eagan RT, Lee RE, Frytak S, Ingle JN, Creagan ET (1980) Combination chemotherapy (CCT) with and without cis-diamminedichloroplatinum(II) plus thoracic radiation therapy (TRT) for limited small cell lung cancer (SCLC). Proc Am Assoc Cancer Res 21: 131 (Abstr 523)

Ettinger DS, Lagakos S (1982) Phase III study of CCNU, cyclophosphamide, adriamycin, vincristine, and VP-16 in small-cell carcinoma of the lung. Cancer 49: 1544–1554

Fox RM, Woods RL, Brodie GN, Tattersall MHN (1980) A randomized study: small cell anaplastic lung cancer treated by combination chemotherapy and adjuvant radiotherapy. Int J Radiat Oncol Biol Phys 6: 1083–1085

Gatzemeier U (1983) Mehrjährige Verlaufsbeobachtung und Ergebnisse nach Chemotherapie (ACO) des kleinzelligen Bronchialkarzinoms. Prax Klin Pneumol 37: 144–147

Ginsberg RJ, Shepherd FA, Evans WK, Feld R, Cooper JD, Ilves R, Todd TRJ, Pearson FG, Waters PF, Baker MA (1983) Chemotherapy followed by adjuvant surgery in the treatment of limited small cell carcinoma (SCLC). 13th International Congress of Chemotherapy, Vienna. Proceedings, part 228, pp 14–18

Goldie JH, Coldman AJ, Gudauskas GA (1982) Rationale for the use of alternating non-cross-resistant chemotherapy. Cancer Treat Rep 66: 439–449

Greco FA, Einhorn LH, Hande KR, Oldham RK (1979) Phase II studies in resistant small cell lung cancer. Proc Am Assoc Cancer Res Am Soc Clin Oncol 20: 28

Green RA, Humphrey E, Close H, Patino ME (1969) Alkylating agents in bronchogenic carcinoma. Am J Med 46: 516–525

Gropp C, Havemann K, Scheuer A (1980) Ectopic hormones in lung cancer patients at diagnosis and during therapy. Cancer 46: 347–354

Hande KR, Oldham RK, Fer MF, Richardson RL, Greco FA (1982) Randomized study of high-dose versus low-dose methotrexate in the treatment of extensive small cell lung cancer. Am J Med 73: 413–419

Hansen HH, Dombernowsky P, Hirsch FR (1978) Staging procedures and prognostic features in small cell anaplastic bronchogenic carcinoma. Semin Oncol 5: 280–287

Hansen HH, Dombernowsky P, Hansen HS, Rørth M (1979) Chemotherapy versus chemotherapy plus radiotherapy in regional small-cell carcinoma of the lung. A randomized trial. Proc Am Assoc Cancer Res 20: 277 (Abstr 1124)

Hansen HH, Dombernowsky P, Hirsch FR, Hansen M, Rygård J (1980) Prophylactic irradiation in bronchogenic small cell anaplastic carcinoma. A comparative trial of localized versus extensive radiotherapy including prophylactic brain irradiation in patients receiving combination chemotherapy. Cancer 46: 279–284

Harper PG, Dally MB, Geddes DM, Spiro SG, Smyth JF, Souhami RL (1982) Epipodophyllotoxin (VP 16-213) in small cell carcinoma of the bronchus resistant to initial combination chemotherapy. Cancer Chemother Pharmacol 7: 179–180

Hattori S, Matsuda M, Tateishi R, Nishihara H, Horai T (1972) Oat-cell carcinoma of the lung. Clinical and morphological studies in relation to its histogenesis. Cancer 30: 1014–1024

Higgins GA (1972) Use of chemotherapy as an adjuvant to surgery for bronchogenic carcinoma. Cancer 30: 1383–1387

Hirsch FR, Matthews MJ, Yesner R (1982) Histopathologic classification of small cell carcinoma of the lung. Comments based on an interobserver examination. Cancer 50: 1360–1366

Hirsch FR, Østerlind K, Hansen HH (1983) The prognostic significance of histopathologic subtyping of small cell carcinoma of the lung according to the classification of the World Health Organization. A study of 375 consecutive patients. Cancer 52: 2144–2150

Høst H (1973) Cyclophosphamide (NSC-26271) as adjuvant to radiotherapy in the treatment of unresectable bronchogenic carcinoma. Cancer Chemother Rep Pt 3, 4 (2): 161–164

Hyde L, Yee J, Wilson R, Patno ME (1965) Cell type and the natural history of lung cancer. JAMA 193: 52–54

Ihde DC, Cohen MH, Bernath AM, Matthews MJ, Bunn PA, Minna JD (1978) Serial fiberoptic bronchoscopy during chemotherapy for small cell carcinoma of the lung. Chest 74: 531–536

Issell B, Rudolph A, Lawson R, Maroun J, Comis R, Hong W, Speer J, Luedke D, Lanzotti V, Hurtubise M, Broder L, Rosenbaum P, White M (1983) The substitution of etoposide for doxorubicin in small cell lung cancer combination chemotherapy. 13th International Congress of Chemotherapy, Vienna, Proceedings, part 248, pp 15–19

Jackson DV, Richards II F, Cooper R, Ferree C, Muss HB, White DR, Spur CL (1977) Prophylactic cranial irradiation in small cell carcinoma of the lung. A randomized study. JAMA 237: 2730–2733

Johnson RE, Brereton HD, Kent CH (1976) Small-cell carcinoma of the lung: attempt to remedy causes of past therapeutic failure. Lancet 2: 289–291

Karrer K, Denck H, Pridun N, Zwintz E, Coop. Group (1983) Combination of early surgery for cure and polychemotherapy in small-cell bronchial carcinoma. 13th International Congress of Chemotherapy, Vienna. Proceedings, Addendum, part 228/52

Kessinger A, Foley JF, Lemon HM (1983) Therapeutic management of small cell lung cancer. Fewer toxic reactions with lower chemotherapeutic drug dosages. JAMA 250: 3188–3191

Komaki R, Cox JD, Whitson W (1981) Risk of brain metastasis from small cell carcinoma of the lung related to length of survival and prophylactic irradiation. Cancer Treat Rep 65: 811–814

Krauss S, Perez C (1980) Treatment of localized undifferentiated small cell lung carcinoma (SCLC) with radiation therapy (RT) with or without combination chemotherapy (CT) with cyclophosphamide (DTIC). In: Hansen HH, Dombernowsky P (eds) Abstracts of the 2nd world conference on lung cancer. Excerpta Medica, Amsterdam, p 157

Livingston RB, Moore TN, Heilbrun L, Bottomley R, Lehane D, Rivkin SE, Thigpen T (1978) Small-cell carcinoma of the lung: combined chemotherapy and radiation. A Southwest Oncology Group study. Ann Intern Med 88: 194–199

Lowenbraun S, Bartolucci A, Smalley RV, Lynn M, Krauss S, Durant JD, Southeastern Cancer Study Group (1979) The superiority of combination chemotherapy over single agent chemotherapy in small cell lung carcinoma. Cancer 44: 406–413

Mason BA, Richter MP, Catalono RB, Creech RB (1982) Upper hemibody and local chest irradiation as consolidation following response to high-dose induction chemotherapy for small cell bronchogenic carcinoma-A pilot study. Cancer Treat Rep 66: 1609–1612

Matthews MJ (1983) Small cell carcinoma of the lung influence of cell subtype in response to therapy. 13th International Congress of Chemotherapy, Vienna. Proceedings, part 228, pp 1–7

Matthews MJ, Rozencweig M, Staquet MJ, Minna JD, Muggia FM (1980) Long-term survivors with small cell carcinoma of the lung. Eur J Cancer 16: 527–531

Maurer LH, Tulloh M, Weiss RB, Blom J, Leone L, Glidewell O, Pajak TF (1980) A randomized combined modality trial in small cell carcinoma of the lung: comparison of combination chemohterapy-radiation therapy versus cyclophosphamide-radiation therapy effects of maintenance chemotherapy and prophylactic whole brain irradiation. Cancer 45: 30–39

McCracken JD, Chen T, White J, Samson M, Stephens R, Coltman CA, Saiki J, Lane M, Bonnet J, McGavran M (1982) Combination chemotherapy, radiotherapy, and BCG immunotherapy in limited small-cell carcinoma of the lung. A Southwest Oncology Group study. Cancer 49: 2252–2258

McMahon LD, Herman TS, Manning MR, Dean JC (1979) Patterns of relapse in patients with small cell carcinoma of the lung treated with adriamycin-cyclophosphamide chemotherapy and radiation therapy. Cancer Treat Rep 63: 359–362

Medical Research Council (1979) Radiotherapy alone or with chemotherapy in the treatment of small-cell carcinoma of the lung. Br J Cancer 40: 1–10

Medical Research Council Lung Cancer Working Party (1981) Radiotherapy alone or with chemotherapy in the treatment of small-cell carcinoma of the lung: the results at 36 months. Br J Cancer 44: 611–617

Mercke C, Arvastsson B, Ewers B, Håkansson L, Kjellen E, Landberg T, Svensson G (1980) Cyclophosphamide, doxorubicin and vincristine combined with radiotherapy in small cell cancer of the bronchus (SBC). In: Hanssen HH, Dombernowsky P (eds) Abstracts of the 2nd world conference on lung cancer. Excerpta Medica, Amsterdam, p 159

Miller AB, Hoogstraten B, Staquet M, Winkler A (1981) Reporting results in cancer treatment. Cancer 47: 207–214

Mountain CF (1974) Surgical therapy in lung cancer: Biologic, physiologic, and technical determinants. Semin Oncol 1: 253–258

Muggia FM, Krezoski SK, Hansen HH (1974) Cell kinetic studies in patients with small cell carcinoma of the lung. Cancer 34: 1683–1690

Muggia FM, Chervu LR (1974) Lung cancer: diagnosis in metastatic sites. Semin Oncol 1: 217–228

Murray N, Klimo P, Goldie J, Hadzic E, Fryer C, Voss N, Gudauskas G (1982) Alternating combination chemotherapy for small cell carcinoma of the lung (SCCL). Proc Am Soc Clin Oncol 1: 151 (Abstr C-587)

Nakhosteen JA, Niederle N (1983) Small cell lung cancer. Serial bronchofiberscopy and photographic documentation − the bridge sign. Chest 83: 12–16

Niederle N (1983) Neue Ansätze in der Diagnostik und Therapie des kleinzelligen Bronchialkarzinoms. Habilitationsschrift, University of Essen

Niederle N, Krischke W, Bremer K, Schmidt CG, Seeber S (1982a) Small-cell bronchogenic carcinoma − primary and relapse therapy with etoposide (VP-16), methotrexate and CCNU. Cancer Treat Rev (Suppl A) 9: 101–105

Niederle N, Krischke W, Schulz U, Schmidt CG, Seeber S (1982b) Untersuchungen zur kurzzeitigen Induktions- und zyklischen Erhaltungstherapie beim inoperablen kleinzelligen Bronchialkarzinom. Klin Wochenschr 60: 829–838

Niederle N, Schütte J, Schmidt CG, Seeber S (1984) Treatment of recurrent small cell lung carcinoma with vindesine and cisplatin. Cancer Treat Rep 68: 791–792

Osoba D, Evans WK, Feld R (1981) VP-16 and cis-platinum (P) combination chemotherapy for relapse in small cell lung cancer (SCLC). Proc Am Assoc Cancer Res Am Soc Clin Oncol 22: 495 (Abstr C-633)

Østerlind K, Sörenson S, Hansen HH, Dombernowsky P, Hirsch FR, Hansen M, Rørth M (1983) Continuous versus alternating combination chemotherapy for advanced small cell carcinoma of the lung. Cancer Res 43: 6085–6089

Peschel PE, Kapp DS, Carter D, Knwolton A (1981) Long term survivors with small cell carcinoma of the lung. Int J Radiat Oncol Biol Phys 7: 1545–1548

Petrovich Z, Mietlowski W, Ohanian M, Cox J (1977) Clinical report on the treatment of locally advanced lung cancer. Cancer 40: 72–77

Pettengill OS, Sorenson GD, Wurster-Hill DH, Curphey TJ, Noll WW, Cate CC, Maurer LH (1980) Isolation and growth characteristics of continuous cell lines from small-cell carcinoma of the lung. Cancer 45: 906–918

Poplin EA, Aisner J, Van Echo DA, Whitacre M, Wiernik PH (1982) CCNU, vincristine, methotrexate, and procarbazine treatment of relapsed small cell lung carcinoma. Cancer Treat Rep 66: 1557–1559

Radice PA, Matthews MJ, Ihde DC, Gazdar AF, Carney DN, Bunn PA, Cohen MH, Fossieck BE, Makuch RW, Minna JD (1982) The clinical behavior of "mixed" small cell/large cell bronchogenic carcinoma compared to "pure" small cell subtypes. Cancer 50: 2894–2902

Rosen ST, Makuch RW, Ihde DC, Matthews MJ, Minna JD, Glatstein E, Bunn PA Jr (1983) Role of prophylactic cranial irradiation in prevention of central nervous system metastases in small cell lung cancer. Am J Med 74: 615–624

Rosenman J, Choi ND (1982) Improved quality of life of patients with small-cell carcinoma of the lung by elective irradiation of the brain. Int J Radiat Oncol Biol Phys 8: 1041–1043

Roszkowski K, Pawlicki M, Slupek A, Traczyk K, Wiatr E, Zych J, Rowinska-Zakrzewska E (1983) VP-16-213 and adriamycin in second line chemotherapy of small cell lung cancer. 13th International Congress of Chemotherapy, Vienna. Proceedings, part 248, pp 47–50

Saugier B, Brunat M, Cordier JF, Gerard JP, Dorsit G, Marchandise JF, Bastidon R, Perol M, Brune J, Galy P (1978) Cancer pulmonaire á "petites cellules". Résultats de l'association adriamycine-vincristine-cyclophosphamide chez 37 patients. Nouv Presse Méd 7: 1357–1361

Schütte J, Niederle N, Krischke W, Seeber S, Schmidt CG (1984) Sequential alternating combination chemotherapy ± radiotherapy in small cell lung cancer (SCLC). Proc Am Assoc Cancer Res 25: 177 (Abstr. 702)

Seeber S, Niederle N, Schilcher RB, Schmidt CG (1980) Adriamycin, Cyclophosphamid and Vincristin ("ACO") beim kleinzelligen Bronchialkarzinom. Verlaufsanalyse und Langzeitergebnisse. Onkologie 3: 5–11

Seydel HG, Creech R, Pagano M, Salazar O, Rubin P, Concannon J, Carbone P, Mohuiddin M, Perez C, Matthews M (1983) Combined modality treatment of regional small cell undifferentiated carcinoma of the lung: a cooperative study of the RTOG and ECOG. Int J Radiat Oncol Biol Phys 9: 1135–1141

Shank B, Natale RB, Hilaris BS, Wittes RE (1981) Treatment of small cell carcinoma of lung with combined high dose mediastinal irradiation, whole brain prophylaxis and chemotherapy. Int J Radiat Oncol Biol Phys 7: 469–475

Shields TW, Higgins GA Jr, Matthews MJ, Keehn RJ (1982a) Surgical resection in the management of small cell carcinoma of the lung. J Thorax Cardiovasc Surg 84: 481–488

Shields TW, Matthews MJ, Higgins GA (1982b) Long term adjuvant chemotherapy in patients after resection of carcinoma of the lung. Third World Conference on Lung Cancer. Tokyo, p 96 (Abstr 117)

Sierocki JS, Hilaris BS, Hopfan S, Martini N, Barton D, Golbey RB, Wittes RE (1979) cis-Dichlorodiammineplatinum(II) and VP-16-213: an active induction regimen for small cell carcinoma of the lung. Cancer Treat Rep 63: 1593–1597

Sierocki JS, Hilaris BS, Hopfan S, Golbey RB, Wittes RE (1980) Small cell carcinoma of the lung. Experience with a six-drug regimen. Cancer 45: 417–422

Smith IE, Sappino AP, Bondy PK, Gilby ED (1981) Long-term survival five years or more after combination chemotherapy and radiotherapy for small cell lung carcinoma. Eur J Cancer Clin Oncol 17: 1249–1253

Sörenson S (1983) Bronchoscopic findings in patients with a complete radiographic regression of small cell bronchogenic carcinoma. Eur J Cancer Clin Oncol 19: 589–595

Souhami RL, Spiro SG, Tobias JS, Geddes DM (1982) Combination chemotherapy and radiotherapy in small cell carcinoma of the bronchus. 3rd World conference on lung cancer, Tokyo (Abstr 213)

Stevens E, Einhorn L, Rohn R (1979) Treatment of limited small cell lung cancer. Proc Am Assoc Cancer Res Am Soc Clin Oncol 20: 435 (Abstr C-599)

Strauchen JA, Egbert BM, Kosek JC, Mackintosh R, Misfeldt DS (1983) Morphologic and clinical determinants of response to therapy in small cell carcinoma of the lung. Cancer 52: 1088–1092

Straus MJ (1974) The growth characteristics of lung cancer and its application to treatment design. Semin Oncol 1: 167–174

Thatcher N, Hunter RD, Jegarajah S, Barber PV, Carroll KB, Wilkinson PM, Crowther D (1982) 11-Week course of sequential methotrexate, thoracic irradiation, and moderate-dose cyclophosphamide for "limited"-stage small-cell bronchogenic carcinoma. A study from the Manchester Lung Tumour Group. Lancet 1: 1040–1043

Valdivieso M, McMurtrey MJ, Farha P, Frazier OH, Barkley HT, Paone JF, Spitzer G, Mountain CF (1982) Increasing importance of adjuvant surgery in the therapy of patients with small cell lung cancer (SCLC). Proc Am Soc Clin Oncol 1: 148 (Abstr C-576)

van Houtte P, Tancini G, de Jager R, Lustman-Maréchal J, Milani F, Bonadonna G, Kenis Y (1979) Small cell carcinoma of the lung: a combined modality treatment. Eur J Cancer 15: 1159–1165

Vermorken JB, Stam J, Van Zandwijk N, Roozendaal KJ, McVie JG, Pinedo HM (1982) Alternating chemotherapy in small cell lung cancer (SCLC). Proc Am Soc Clin Oncol 1: 149 (Abstr C-580)

Vincent RG, Wilson HE, Lane WW, Chen TY, Raza S, Gutierrez AC, Caracandas JE (1981) Progress in the chemotherapy of small cell carcinoma of the lung. Cancer 47: 229–235

Weiss W, Boucot KR, Cooper DA (1970) The histopathology of bronchogenic carcinoma and its relation to growth rate, metastasis, and prognosis. Cancer 26: 965–970

Whang-Peng J, Kao-Shan CS, Lee EC, Bunn PA, Carney DN, Gazdar AF, Minna JD (1982) Specific chromosome defect associated with human small-cell lung cancer: deletion 3p(14-23). Science 215: 181–182

Wolf J, Patno ME, Roswit B, D'Esopo N (1966) Controlled study of survival of patients with clinically inoperable lung cancer treated with radiation therapy. Am J Med 40: 360–367

Woods RL, Levi JA (1983) Cis-platinum and VP16-213 treatment of small cell lung cancer (SCLC) – a randomised study of duration of therapy. Third World Conference on Lung Cancer. Tokyo (Abstr 217)

Young JA, Dillman RO, Seagren SL, Taetle R, Rentschler RE, Lea JW Jr, Lehar TJ, Green MR, Stanton W, Mendelsohn J, Royston I (1982) Non-cross-resistant chemotherapy and consolidation radiotherapy for small cell carcinoma of the lung. Cancer Treat Rep 66: 1399–1401

Zelen M (1973) Keynote address on biostatistics and data retrieval. Cancer Chemother Rep Pt. 3, 4(2): 31–42

Intensive Chemotherapy with Autologous Bone Marrow Transplantation in Small Cell Carcinoma of the Lung

P. G. Harper and R. L. Souhami

Department of Medical Oncology, Guy's Hospital,
St. Thomas Street, London SE1, 9RT, England

Introduction

In spite of improvements in response rate with combination chemotherapy, median survival is poor in both limited and extensive disease. The best figures for median survival are 18 months in limited disease and 12 months in extensive disease. In large trials, where less selection may have taken place, the survival figures are worse, usually something like 15 months and 9 months, respectively. The adoption of the strategy of using "non-cross-resistant" chemotherapy has made little impact on these dismal figures. Approximately 20% of patients with limited disease will live 2 years from diagnosis, representing 5% of all patients. Furthermore, relapses are frequently seen beyond 2 years.

These disappointing results have led to an investigation of very-high-dose chemotherapy in an attempt to increase both response and the proportion of long-term survivors.

Evidence for a Dose/Response Relationship for Cytotoxic Drugs

The use of cytotoxic agents in very high doses has theoretical attractions in a relatively sensitive tumour such as SCLC. There is now a considerable body of data suggesting that chemotherapy dose is of major importance in determining outcome in sensitive tumours (Frei and Canellos 1980). However, many different pharmacological and biological considerations must be taken into account in devising high-dose regimens for cancer:

a) Is There Evidence that a Steep Dose/Response Relationship Exists for Each Drug to be Used? If the intention is to increase *the dose* administered on each occasion (rather than to increase the *frequency* of administration of conventional doses) then it follows that this will only be useful if a dose/response relationship exists for the drug.

For alkylating agents and nitrosoureas such a relationship has been shown experimentally. In rapidly growing tumours cultures in vitro or transplanted to animals a dose/response relationship with alkylating agents was demonstrated by Bruce et al. in 1966. In studies of the drug sensitivity of xenografts of human tumours in immune-suppressed mice, Shorthouse et al. (1982) showed that there was a relationship between growth delay and dose of some cytotoxic drugs. With SCLC xenografts there was a linear relationship between growth delay and dose of cyclophosphamide, procarbazine, and CCNU. No dose/response relationship was found with methotrexate.

Clinical studies have supported these findings. Increased responses to high-dose melphalan have been found in melanoma (McElwain et al. 1982). High-dose BCNU appears to

Recent Results in Cancer Research. Vol. 97
© Springer-Verlag Berlin · Heidelberg 1985

produce responses in glioma (Hochberg et al. 1981), but at the expense of considerable toxicity (see below). Cyclophosphamide has been given in very high doses to patients with ovarian and other cancers (Buckner et al. 1974; Rudolph et al. 1972), with a possibly increased response rate in ovarian cancer. In SCBC high-dose cyclophosphamide produces a greatly increased response rate compared with the conventional dose (Souhami et al. 1982).

Only a few other drugs have been used at high doses. There is some evidence that VP16-213 in doses of $1.2-2.4$ g/m^2 may be associated with a higher response rate in tumours such as SCLC, glioblastoma, and Hodgkin's disease (Wolff et al. 1982), but we have found the response to doses of 1 g/m^2 in previously treated gliomas to be disappointing (G Finn, R. L. Souhami and D. G. Thomas, unpublished observations). High-dose methotrexate is probably associated with an increased response rate in osteosarcoma (Jaffe et al. 1972; Rosen and Nirenberg 1982) and lymphoma (Djerassi and Kim 1976), but in medulloblastoma the effect on response rates is less convincing (Mooney et al. 1983). In SCLC there appears to be no advantage in increasing doses of methotrexate. Hande et al. (1982) Compared 6 g/m^2 with 20 mg/m^2 in combination with cyclophosphamide, adriamycin, and vincristine, and found no improvement in response rate or survival for the high-dose group. Arnold et al. (1984) found no advantage in a methotrexate dose of 1 g/m^2 compared with 200 mg/m^2. For many other drugs, e.g., adriamycin, there is no evidence that an increase beyond the usually accepted dose range will lead to an increased response rate. With some agents, for example vincristine, unacceptable toxicity makes dose escalation unsafe (see below) while in other cases adequate evaluation has not been undertaken.

There are additionally many possible factors which may limit the value of dose escalation. (i) In slowly growing insensitive tumours the increase in tumour kill may be clinically insignificant but the price paid in toxicity may be high. Even in sensitive tumours the growth fraction may be small and this will limit the response to a single dose, particularly of cycle-active agents. (ii) Large tumours may have areas of very poor vascularity allowing inadequate drug penetration yet still containing viable cells. (iii) Inherent drug resistance may also be related to tumour size (Goldie and Coldman 1979). Dosage schedules which are based on data derived from rapidly dividing experimental tumours growing as ascites or leukaemia, or in vitro, may not be applicable to the treatment of human cancer.

b) What is the Optimum Dosage Schedule for Each Drug? The optimum method of administration should be determined for each agent, that is whether it is better to administer the higher dose by a single bolus, by infusion, or by repeated boluses over a few days. In the case of melphalan the drug is rapidly cleared from the circulation following a high dose given as a bolus (McElwain et al. 1979), but it is not known whether this is a more effective way of giving the drug than repeated injections or infusion. The rapid clearance means that harvested bone marrow can be returned to the patient without the need for cryopreservation, which is expensive and time-consuming and limits the general applicability of the autografting techniques.

With cyclophosphamide the drug must be converted by the liver into 4-hydroxy-cyclophosphamide before being active. We have recently shown that when 50 mg/kg is given on each of 4 successive days (total dose 200 mg/kg) the t$^1/_2$ (half-life) is progressively shortened (Graham et al. 1983). The mechanism is unknown but there is no increase in urinary excretion of the drug. This indicates that the ability of the liver to convert the drug is not exceeded by this dose schedule. Similarly, when the same total dose is given in four injections at 3-h intervals, the ability of the liver to convert the drug does not seem to be exceeded (data quoted by Kaye 1982). It is not known whether the two schedules have an equal antitumour effect. In the study of Ettinger et al. (1973) cyclophosphamide 60 mg/kg was given either on days 1 and 2 or on days 1 and 8. There was no difference in the response rate (70%) but the toxicity was somewhat less with the day 1 and 8 schedule.

At the present time the dose schedules which have been used in very-high-dose chemotherapy have understandably been arrived at as a result of the experiences of the investigator and of the availability of facilities such as cryopreservation, rather than being based on response rates and pharmacological principles.

c) What are the limiting toxicities? Increasing the dose of a cytotoxic drug increases its toxicity. In the case of the cycle-phase-specific agent methotrexate the toxicity is minimal, provided that the drug action is terminated at approximately 24 h with folinic acid. With alkylating agents the toxicity is primarily haematological and can be mitigated to some extent by autologous bone marrow transplantation (ABMT). The value of ABMT in lessening the period of aplasia has been demonstrated for melphalan (McElwain et al. 1979). We have attempted to assess the value of ABMT in limiting the period of aplasia when cyclophosphamide is given at a dose of 50 mg/kg on 4 successive days. With increasing delay in the return of cryopreserved bone marrow there did not appear to be any prolongation of the period of neutropenia (Table 1). However, our recent experience of repeating the high-dose treatment after an interval of 4 weeks suggests that marrow reserves are reduced following the first treatment (Goldstone et al. 1983), and in this situation ABMT may be essential even though the dose of drug has not been changed. Clearly the need for ABMT has to be defined for each chemotherapy schedule. Smith et al. (1983) have shown that ABMT is not necessary when 7 g/m^2 cyclophosphamide is given over 12 h. ABMT is almost certainly necessary following very-high-dose BCNU.

Other toxicities may prevent further dose increases, however. In the case of cyclophosphamide, haemorrhagic cystitis is a severe and occasionally fatal complication. The major cause of urothelial toxicity appears to be acrolein, which is formed when cyclophosphamide is metabolised (Cox et al. 1979). Attempts to avoid this complication by vigorous hydration regimens and bladder irrigation have had limited success. The development of 2-mercaptoethane sulphonate (Bryant et al. 1980) has been a major advance in preventing haemorrhagic cystitis, although careful attention to details of dosage and timing are essential.

Gut toxicity is not a problem with cyclophosphamide in doses up to 200 mg/kg, but is severe with BCNU and melphalan. Mucositis is the major side-effect with VP16-213 at doses over 1.8 g/m^2, but neurotoxicity has not been a major problem (Wolff et al. 1982).

Carditis is a reported complication of cyclophosphamide treatment. In the series of Buckner et al. (1972), five patients receiving 120 mg/kg developed ECG changes and one

Table 1. Duration of severe neutropenia[a] related to day of autograft post chemotherapy

Days after chemotherapy	Total dose of cyclophosphate (mg/kg)	Duration of neutropenia (days ± SD)
2	160	10 ± 2.45
	200	12.75 ± 2.4
4	200	12.75 ± 3.4
6	200	13 ± 2.16

[a] Neutrophil count $< 0.5 \times 10^9$/l

patient receiving 60 mg/kg on 4 consecutive days developed fatal haemorrhagic carditis. This complication has also been seen in the BACT protocol (Appelbaum et al. 1978). With cyclophosphamide 1.6 g/m^2 on each of 4 successive days with BCNU, cytosine arabinoside and 6-thioguanine, we have not seen ECG abnormalities or carditis in 65 patients treated with high-dose cyclophosphamide (200 mg/kg) alone in a schedule including 2-mercaptoethane sulphonate as prophylaxis against haemorrhagic cystitis. The lack of cardiac toxicity, even in patients who have been treated with 200 mg/kg on two occasions, might possibly be due to a protective effect of 2-mercaptoethane sulphonate on the heart as well as on the bladder.

Pulmonary toxicity is a well-recognised and sometimes fatal complication of nitrosoureas and total-body irradiation, and is occasionally seen with cyclophosphamide. We have not seen interstitial pneumonitis in any patient treated with high-dose cyclophosphamide, but dense fibrosis has developed in 75% of patients at the site of mediastinal irradiation with 40 Gy in 20 fractions (Trask et al. 1984).

d) What is the Value of Combining Drugs in High Doses? The term high-dose chemotherapy is also used to describe drug *combination* regimens in which the total dose of one or more drugs is increased to a variable extent, and this is combined with other drugs at conventional, or slightly increased, doses. Examples are the BACT protocol used for Burkitt's lymphoma (Appelbaum 1978): (BCNU 200 mg/m^2 day 1, cyclophosphamide 1.6 g/m^2 on days 2–5, cytosine arabinoside 200 mg/m^2 on days 2–5, and 6 thioguanine 200 mg/m^2 on days 2–5) and the protocol of cyclophosphamide 1.5 g/m^2 on days 1–3, VP16 200 mg/m^2 on days 1–3, vincristine 1.5 mg/m^2 on days 1 and 3, and adriamycin 80 mg/m^2 on day 1, as used for SCLC by Farha et al. (1983b). In both these protocols cyclophosphamide doses are 3–4 times those used conventionally, and the other drugs are used in full or twice-normal dosage. These regimens do *not* address themselves to the question of whether useful dose/response relationships exist. They *assume* that this is the case and that several drugs combined at supranormal doses will be more effective than the same drugs at normal doses. As discussed below the problem is that the relative contribution of each drug is hard to assess, and it is possible that extra and unexpected toxicity may occur without therapeutic benefit. This is seen particularly when high doses of drugs are combined with toal-body irradiation (TBI) (Douer et al. 1981).

The aim of high-dose treatment in solid tumours is to treat the cancer, and marrow aplasia is an unwanted effect. When devising schedules using multiple agents it is important to be sure that each component contributes to this objective, otherwise needless toxicity will be inflicted.

e) Is the Approach Practicable for Patients with SCLC? Patients with SCLC are usually middle-aged or elderly. Most have been life-long heavy smokers. They therefore have an increased incidence of associated chronic bronchitis and emphysema, peripheral vascular disease, and coronary artery disease. Two-thirds of the patients have extensive disease at presentation, and half of our patients have a Karnovsky performance status of 70 or lower at presentation. Cyclical intermittent chemotherapy is toxic and debilitating for many patients, and the toxicity of "late intensification" with high-dose chemotherapy is likely to be considerable.

Results of High-Dose Treatment

As early as 1972 attempts were made to escalate the dose of chemotherapy with previously stored bone marrow used to ameliorate the haematological toxicity. Buckner et al. (1972)

used cyclophosphamide in doses ranging from 60 mg/kg to 240 mg/kg in 26 patients with a variety of resistant tumours, including a single case of oat cell carcinoma of the bronchus and three adenocarcinomas of the lung. Unfortunately, no responses were seen in this group. Although median WBC nadirs varied from 500×10^8/litre at the 60 mg/kg dose to 120×10^8/litre at the 120 mg/kg dose, the duration of neutropenia was short (approx. 12 days) with no benefit from autologous bone marrow support.

Few studies were reported in solid tumours until the late 1970s, when Gale et al. (1979) reported a single case of SCLC where bone marrow had been cryopreserved at diagnosis followed by conventional chemotherapy with six agents. At relapse he was treated with local radiation to metastatic sites, in addition to chemotherapy with high-dose cyclophosphamide, adriamycin, methotrexate, and TBI. A complete remission was obtained.

There have more recently been several reports of high-dose chemotherapy in SCLC using a variety of different treatment strategies.

a) Intensive Drug Therapy on Relapse. Various studies have been reported in which high-dose combination chemotherapy has been given to patients in relapse after multiple-drug therapy. Gale et al. (1979) used a single course of vinblastine 1.0 mg/kg, adriamycin 120−200 mg, and cyclophosphamide 100−200 mg/kg in ten patients, with the addition of TBI (8 Gy) in four patients, achieving a complete remission in the only patient with oat cell carcinoma, who survived a further 4 months. The median time to recovery of neutrophils to 0.5×10^9/litre was 19 days. Four of nine other patients whose treatment included adriamycin died of cardiac failure. Douer et al. (1981), in a series of 14 previously treated patients, included 1 patient with small cell lung cancer who also entered complete remission with cyclophosphamide, vinblastine, and TBI. The remission lasted 2 months and the patient survived a total of 122 days from the intensive therapy. In a series of 19 patients Spitzer et al. (1980) included 9 patients with small cell lung cancer, 8 of whom had relapsed following very aggressive conventional therapy of 4−20 months' duration. Intensive chemotherapy was with cyclophosphamide 4.5 g/m^2 over 3 days, BCNU 300 mg/m^2, and VP16 600 mg/m^2 over 4 days. Autologous bone marrow was reinfused on the 6th day. Six of the eight evaluable patients achieved a response, including one complete response, lasting from 4+ to 55+ weeks. Pico et al. (1983) treated ten patients with limited-stage small cell carcinoma in which "planned" bone marrow harvest and cryopreservation took place prior to the second course of conventional therapy, which comprised adriamycin, VP16, cisplatinum, *Corynebacterium parvum,* and radiotherapy. High-dose therapy was given at the time of relapse. At the time of the report four patients had relapsed and intensive chemotherapy was given, which consisted in BCNU 300 mg/m^2 day 1, procarbazine 200 mg/m^2 on days 1−4 and melphalan 140 mg/m^2 on day 5. Following autologous marrow reinfusion neutrophil recovery ($< 500 \times 10^8$/litre) took 6−11 days. Two complete and two partial responses were obtained, with survival of 6, 16, and 27 weeks and one early death at 15 days from cardiorespiratory failure.

These studies have demonstrated that intensive treatment can achieve remissions (albeit of short duration) at the time of relapse when further conventional drug treatment seldom produces a response, still less a complete response.

b) Intensive Drug Therapy as Initial Treatment. Following the demonstration that high-dose treatment produced responses in the difficult clinical situation of relapsed drug-resistant tumours, the next step was to use this approach under more favourable clinical circumstances. The strategies have been to use high-dose chemotherapy either at presentation (using single drugs or combination therapy) *or* as a policy of "late intensification" when tumour bulk is at its minimum.

In 1979 we embarked on a pilot study of a single course of very-high-dose cyclophosphamide (200 mg/kg over 4 days) with ABMT using cryopreserved marrow and 2-mercaptoethane sodium sulphonate to prevent haemorrhagic cystitis. Full restaging was carried out 4 weeks after high-dose therapy. Subsequently radiotherapy (40 Gy in 20 fractions in 28 days) was given with the field covering the mediastinum and site of the original primary disease. No further treatment was given until relapse.

We felt the structure of such a "simple" study would enable us to answer the following questions:

1) Was such an approach feasible in this predominantly middle-aged and elderly group? The toxicity of a single-agent study was likely to be tolerable.
2) Would responses to very-high-dose cyclophosphamide be superior to those we would expect with conventional doses?
3) Would a single course of treatment result in a relapse-free survival comparable to that achieved with conventional therapy? If so this single course of treatment might be preferable in morbidity to repeated cyclical therapy.
4) On relapse, would the tumour still be fully sensitive to combination therapy?

Our results (Souhami et al. 1982, 1983) in 25 patients showed that there was a high overall response rate to the single cycle of therapy (84%), with 14/25 (56%) achieving a bronchoscopically proven complete response. The main toxicity was haematological, but it was manageable. There were no treatment-related deaths. No carditis occurred after irradiation; however, as mentioned above, local pulmonary fibrosis was striking and raises the possibility that high-dose cyclophosphamide can sensitise the lung to radiotherapy. The median disease-free interval was 49 weeks, with a median survival of 66 weeks. Five patients (20%) were disease-free at 2 years and three patients at 3 years. Patients in remission led a life which was normal in every way. The complete and partial response rate to second-line therapy (*cis*-platinum and VP16-213) is 60%, indicating no cross resistance.

We are now investigating whether two cycles of high-dose cyclophosphamide followed by radiotherapy result in a higher response rate and longer disease-free period. It is too early to assess response rate and duration, but it is already clear that the haematological toxicity is greater following the second course of therapy, with a delay in recovery of both neutrophils and platelets (Goldstone et al. 1983).

When ABMT is used in untreated patients there is a risk that the harvested marrow may be contaminated with tumor cells. In our cases the marrow is examined carefully for tumour before being reinfused, but if the tumour cells are not in recognisable clumps a small degree of contamination might not be recognised. We are now examining the marrow using monoclonal antibodies reactive with SCLC lines in an attempt to increase the sensitivity of detection. However, it is not known whether SCLC will withstand freezing and thawing *and* there is likely to be more disease left in the patient after treatment than is reinfused in the marrow. Using thoracic CT scanning we have shown that even in patients with a bronchoscopic complete response, mediastinal abnormalities remained. It does not appear likely that small degrees of contamination will influence the results of treatment at present. It could do so if marrow harvested at presentation is divided and reinfused after repeated high-dose administration (as in our present study). This problem may be avoided if the first treatment can be carried out without marrow and the harvest performed after recovery, or if reliable methods can be developed for the elimination of contaminating tumour cells.

The only other study of high-dose *single-agent* therapy as initial treatment has been carried out by Johnson et al. (1983). Thirteen patients with extensive small cell carcinoma were treated with VP16 1,200 mg/m^2 over 3 days for two courses. Previous studies had shown that bone marrow support was not required (Wolff et al. 1982). Eight of ten patients responded (80%), with four patients attaining complete remission. Median survival for complete responders was 11 months. Further therapy with conventional doses of cyclophosphamide, adriamycin, and vincristine produced no further responses. There were three early treatment-related deaths. Fever during neutropenia occurred in 11/26 courses, with proven bacterial infection in 5, and mild to moderate mucositis was common. There does therefore appear to be a dose/response relationship with VP16 in SCLC, since previous studies had indicated overall responses in such patients to be 40%−50% with standard doses.

Farha et al. (1983a, b) used a combination regimen with autologous marrow support in 14 previously untreated patients. Cyclophosphamide (1.5 g/m^2 days 1−3), VP16 (200 mg/m^2 days 1−3) and vincristine (1.5 mg/m^2 day 1 and 3) were used in all, but adriamycin (80 mg/m^2 day 1), used additionally in the first 6 cases, was abandoned because of cardiac arrythmias.

In all, 5 of 8 patient (62%) with limited disease and 2 of 5 (40%) patients with extensive disease attained complete remission. Further maintenance therapy was then given with the same drugs in lower dosage. Prophylactic cranial radiation was given to all, and chest radiotherapy to all partial responders and to half the complete responders. Overall median survival was 54 weeks with complete responders having a median remission of 40 weeks and median survival of 64 weeks. There was no outstanding systemic toxicity. Haematological toxicity was compared to a group of patients treated with the same drugs on a less aggressive schedule who did not have ABMT and the authors concluded that ABMT had shortened the period of aplasia. Overall response rate and remission duration were not significantly different to conventional studies.

c) Intensive Drug Therapy as Late Intensification. The strategy of late intensification relies on the hypothesis that the tumour is more sensitive when it is small. The assessment of results of the approach is very difficult, however. First, response rates can only be assessed in the patients whose tumours remain after initial treatment − a group who are unlikely to benefit. Secondly, in complete responders the value of the late intensification is only measurable in terms of survival. Unless this effect is dramatic it can only be shown by a randomised prospective study comparing late intensification with conventional therapy. Thirdly, in the design of the study a decision must be made as to whether effective drugs should be reserved for use only in the intensification regimen. This raises the question of whether there is a genuine lack of cross resistance between one cytotoxic agent and another. Finally, only a proportion of patients starting the protocol will be eligible for late intensification and these will be a selected population, thus limiting the value of the approach.

Smith et al. (1983) used high doses cyclophosphamide (7 g/m^2), an agent not included in their initial treatment schedule, as a method of intensifying treatment following a course of "conventional therapy" in 27 patients. Twelve patients (45%) attaining complete remission with conventional therapy had a median disease-free survival of 33 weeks and overall survival of 52 weeks. Of ten partial responders, and five patients with progressive disease, late intensification achieved a CR in four and a PR in nine. Median duration of remission was short (10 weeks) and median survival 36 weeks. No overall benefit in survival appears to have been gained compared with conventional therapy.

The Brussels group (Klastersky et al. 1982; Sculier et al. 1983) has reported two studies with late intensification. In their first study two courses of *cis*-platinum, adriamycin, and VP16 (CAV) were given to 36 patients (14 with limited, 22 with extensive disease). Late intensification was carried out in 13 patients, and consisted in either double the conventional dose given once or twice, or three times the conventional dose given once. Of seven patients with PR intensification produced CR·in only two, and a further patient with progressive disease achieved PR. There were two toxic deaths. Mucositis was severe in the group treated with triple the dose of induction therapy, and no survival advantage was apparent. Thus CAV did not appear to be the optimal therapy for further intensification. In their second study induction therapy was either adriamycin, VP16, and cyclophosphamide (Endoxana) (AVE) (7 patients) or the same combination with the addition of *cis*-platinum (CAVE) (1 patient) for three courses. Late intensification was given in the form of adriamycin, cyclophosphamide, and VP16 in three cases. Because of mucositis the adriamycin was abandoned and the other two drugs were used alone in the next 11 cases, with the dose of cyclophosphamide escalated from 100 mg/kg to 200 mg/kg and that of VP16 from 750 mg/m^2 to 1 g/m^2. Four of eight partial responders showed an additional response to the intensification. The usefulness of ABMT was not determined.

Ihde et al. (1983) recently reported a study in extensive disease in which the induction therapy was their conventional therapy (cyclophosphamide, methotrexate, and CCNU for 6 weeks, followed by vincristine, adriamycin, and procarbazine for 6 weeks), so that a comparison can be made with their previous studies. Autologous marrow was collected in the group attaining CR or PR, with good performance status, who were bone marrow negative. Intensification consisted of radiotherapy (20 Gy in 5 days) to all sites of initial tumour and chemotherapy (cyclophosphamide 60 mg/kg × 2 days, VP16 200 mg/m^2 × 3 days) with ABMT. Following prophylactic cranial irradiation no further treatment was given. Due to lack of initial tumour regression and poor performance score only 10 of 29 patients were eligible. Two refused treatment and, of the 8 treated, 3 were in CR at the time of intensification but relapsed at 4, 8, and 15 months. Of the 5 with PR, only 1 had further tumour regression and this was of short duration (3 months). The authors have drawn attention to the lack of improvement over their previous conventional studies and do not feel that the high-dose therapy in this study is likely to lead to improvement.

In the final study from the MD Anderson Hospital, Farha et al. (1983b) have reported on 21 patients with limited disease. Induction consisted in three courses of vincristine, adriamycin, and ifosfamide in 10 patients and vincristine, adriamycin, cyclophosphamide, and VP16 in 11 patients. Intensification with cyclophosphamide, VP16, and vincristine was as used in their previous studies, with 11 patients receiving methotrexate in addition (1.5 g/m^2 with folinic acid rescue). Initial therapy gave a 91% response rate with 9/21 (43%) CR. Late intensive treatment caused five additional patients to enter CR. After late intensive treatment prophylactic cranial irradiation and thoracic irradiation was administered to all patients in PR and half those in CR, followed by four further courses of chemotherapy. Survival for patients in CR is 16+ months and for those in PR 10+ months. Myelosuppression was prolonged, with a considerable number of infective episodes.

Banham et al. (1982) used cyclophosphamide, adriamycin, vincristine, and prednisone (CHOP) as their initial induction regimen, with a poor response rate. A change to cyclophosphamide, adriamycin, and VP16, with vincristine and methotrexate on day 10 has increased the response rate. Of 67 patients treated, 20 (55%) of 36 patients with limited disease, and 5 (16%) of 31 with extensive disease, achieved CR. High-dose intensification was achieved with an 18-h treatment consisting of cyclophosphamide 200 mg/kg and VP16 200 mg/m^2. This has been carried out in 13 of the 25 patients who achieved CR and in 7 of

23 patients with PR following conventional therapy. ABMT was used. The long-term results are awaited.

Stewart et al. (1983) reported on ten patients who have been treated with a late intensification schedule (cyclophosphamide + BCNU) including TBI after a variety of induction regimens. Only three of seven evaluable patients showed an additional response. Two patients died of interstitial pneumonitis, one of radiation-associated liver change, and six of progressive disease.

In summary, late intensification has not proved to be dramatically beneficial, and its evaluation in terms of response is difficult. With some regimens toxicity and treatment-related morbidity, principally infection and mucositis, have been severe. Most reports have *not* stated the days spent as an in-patient, but this must be significant in the light of the poor results. Haematological toxicity has on the whole not been a problem, with satisfactory marrow recovery in all. Whether autologous marrow support is needed had not been determined for most schedules. Certainly, there is no evidence that this is a useful technique for patients with extensive disease or poor performance score. The only group for whom there may be some benefit is that of patients with limited disease who have achieved a complete response with conventional therapy. Many of the studies are difficult to interpret because of inconsistency in primary therapy and changes in late intensification schedules, resulting in very small groups of heterogeneous patients.

Conclusion

In summary, there is evidence that in SCLC a dose/response relationship exists for some drugs. Responses to high-dose treatment have been seen in patients who have drug-resistant tumours. Late intensification has not yet proved beneficial but interpretation of the data is very difficult. Further systematic studies on high-dose induction regimens are needed.

References

Appelbaum FR, Deisseroth AB, Graw RG, Herzig GP, Levine AS, McGrath IT, Pizzo PA, Poplack DG, Ziegler JL (1978) Prolonged complete remission following high dose chemotherapy of Burkitt's lymphoma in relapse. Cancer 41: 1059–1063

Arnold AM, Williams CJ, Mead GM, Buchanan RB, Green JA, Macbeth FM, Whitehouse JM (1984) Combination chemotherapy using high or low dose Methotrexate for small cell carcinoma of the lung: a randomised trial. Med Oncol Tumour Pharmocother 1: 9–14

Banham S, Burnett A, Stevenson R, Cunningham D, Kaye S, Ahmenzai S, Soukop M (1982) A pilot study of combination chemotherapy with late dose intensification and autologous marrow rescue in small cell bronchial carcinoma. Br J Cancer 46: 486–488

Bruce WB, Meeker BE, Valeriote FA (1966) Comparison of the sensitivity of normal haemopoietic and transplanted lymphoma colony forming cells to chemotherapeutic agents administered in vivo. J Natl Cancer Inst 36: 233–245

Bryant B, Jarman M, Ford H, Smith IE (1980) Prevention of ifosfamide induced urothelial toxicity with 2-mercaptoethane sulphonate sodium (mesna) in patients with advanced carcinoma. Lancet 2: 657–659

Buckner CD, Rudolph RH, Fefer A, Clift RA, Epstein RB, Funk DD, Neiman PE, Slichter SJ, Storb R, Thomas ED (1972) High dose cyclophosphamide therapy for malignant disease. Toxicity, tumor response and the effects of stored autologous marrow. Cancer 29: 357–365

Buckner CD, Briggs R, Clift RA, Fefer A, Funk DD, Glucksberg H, Neiman PE, Storb E, Thomas ED (1974) Intermittent high-dose cyclophosphamide treatment of stage II ovarian carcinoma. Cancer Chemother Rep 58: 697–703

Cox PJ (1979) Cyclophosphamide cystitis – identification of acrolein as the causative agent. Biochem Pharmacol 28: 2045–2049

Djerassi I, Kim JS (1976) Methotrexate and citrovorum factor rescue in the management of childhood lymphosarcoma and reticulum cell sarcoma (non-Hodgkin's lymphoma). Cancer 38: 1043–1051

Douer D, Champlin RE, Ho WG, Sarna GP, Wells JH, Graze PR, Cline MJ, Gale RP (1981) High dose combined-modality therapy and autologous bone marrow transplantation in resistant cancer. Am J Med 71: 973–976

Eagan RT, Carr DT, Frytak S, Rubin J, Lee RE (1976) VP16-213 versus polychemotherapy in patients with advanced small cell lung cancer. Cancer Treat Rep 60: 949–952

Ettinger DS, Karp JE, Abeloff MD, Burke PJ, Braine HG (1978) Intermittent high-dose cyclophosphamide chemotherapy for small cell carcinoma of the lung. Cancer Treat Rep 62: 413–424

Farha P, Spitzer G, Valdivieso M (1983a) High dose intensification with autologous bone marrow transplantation in limited disease small cell lung cancer. In: Proceedings of UCLA symposium on recent advances in bone marrow transplantation, Salt Lake City, Feb 13–18, 1983, Abstract 0155

Farha P, Spitzer G, Valdivieso M, Dicke KA, Sander A, Dhingra HM, Minhaar G, Vellekoop L, Verma DS, Umsawasdi MD, Chiuten D (1983b) High-dose chemotherapy and autologous bone marrow transplantation for the treatment of small cell lung carcinoma. Cancer 52: 1351–1355

Frei IE, Canellos GP (1980) Dose: A critical factor in cancer chemotherapy. Am J Med 69: 585–594

Gale R, Graze PR, Wells J, Ho W, Hershko C, Lowenberg B, Felgs S, Cline MJ (1979) Autologous bone marrow transplantation in patients with cancer. Exp Haematol [Suppl 5] 7: 351–359

Goldie JH, Coldman AJ (1979) A mathematic model correlating the drug sensitivity of tumors to the spontaneous mutation rate. Cancer Treat Rep 63: 1727–1733

Goldstone AH, Harper PG, Linch DC, Souhami RL, Anderson CC, Moss FM, Martin PJ, Richards JDM (1983) Patterns of haematological recovery in autologous bone marrow transplantation for small cell carcinoma of bronchus, carcinoma of the ovary and acute leukaemia and lymphoma. Exp Haematol [Suppl 13] 11: 164–167

Graham MI, Shaw IC, Souhami RL, Sidau B, Harper PG, McLean AEM (1983) Decreased plasma half-life of cyclophosphamide during repeated high-dose administration. Cancer Chemother Pharmacol 10: 192–193

Hande KR, Oldham RK, Fer MF, Richardson RL, Greco FA (1982) Randomised study of high-dose versus low-dose methotrexate in the treatment of extensive small cell lung cancer. Am J Med 73: 413–419

Hochberg FH, Parker LM, Takvorian T, Canellos GP, Zervas NT (1981) High-dose BCNU with autologous bone marrow rescue for glioblastoma multiforme. J Neurosurg 54: 455–460

Ihde DC, Lichter AS, Deisseroth AB, Bunn PA, Carney DN, Cohen MH, Makuch RW, Johnston-Early A, Minna J (1983) Late intensive combined-modality therapy with autologous bone marrow infusion in extensive stage small cell lung cancer. Am Assoc Clin Oncol 19(C7747: 198

Jaffe N (1972) Recent advances in the chemotherapy of metastatic osteosarcoma. Cancer 30: 1627–1631

Johnson DH, Wolff SN, Hande KR, Hainsworth JD, Fer MF, Greco FA (1983) High-dose VP16-312 treatment of extensive small cell lung cancer. Am Assoc Clin Oncol 24: C754

Kaye SB (1982) Intensive chemotherapy for solid tumours – current clinical applications. Cancer Chemother Pharmacol 9: 127–133

Klastersky J, Nicaise C, Longeval E, Strykmans P (1982) cis-Platin, adriamycin, etoposide (CAV) for remission induction of small-cell bronchogenic carcinoma. Cancer 50: 652–658

McElwain TJ, Hedley DW, Burton G, Clink HM, Gordon MY, Jarman M, Juttner CA, Millar JL, Milsted RAV, Prentice G, Smith IE, Spence D, Woods M (1979) Marrow autotransplantation accelerates haematological recovery in patients with malignant melanoma treated with high-dose melphalan. Br J Cancer 40: 72−80

Mooney C, Souhami RL, Pritchard J (1983) Recurrent medulloblastoma: Lack of response to high-dose methotrexate. Cancer Chemother Pharmacol 10: 135−136

Pico JL, Beaujean F, Debre M, Carde P, Le Cheumlier T, Haat M (1983) High-dose chemotherapy with autologous bone marrow transplantation in small cell carcinoma of the lung in relapse. Proc Am Soc Clin Oncol 19: (C806): 206

Pritchard J, McElwain TJ, Graham-Pole G (1982) High dose melphalan with autologous marrow for treatment of advanced neuroblastoma. Br J Cancer 45: 86−88

Rosen G, Nirenberg A (1982) Chemotherapy for osteosarcoma: An investigative method, not a recipe. Cancer Treat Rep 66: 1687−1697

Sculier JP, Klastersky J, Strykckmans P, Weerts D (1983) Late intensification in small cell lung cancer: Result of a pilot study. Proc Am Assoc Canc Res 24: 1052

Shorthouse AJ, Jones JM, Steel GG, Peckham MJ (1982) Experimental combination and single-agent chemotherapy in human lung-tumour xenografts. Br J Cancer 46: 35−45

Smith IE, Evans BD, Harland SJ (1983) High-dose cyclophosphamide (7 g/m^2) ± autologous bone marrow rescue after conventional chemotherapy in patients with small cell lung cancer. Am Assoc Clin Oncol 2: C726

Souhami RL, Harper PG, Linch DC, Trask C, Goldstone AH, Tobias J, Spiro SG, Geddes DM, Richards JDM (1982) High-dose cyclophosphamide with autologous marrow transplantation as initial treatment of small cell carcinoma of the bronchus. Cancer Chemother Pharmacol 8: 31−34

Souhami RL, Harper PG, Linch D, Trask C, Goldstone AH, Tobias JS, Spiro SG, Geddes DM, Richards JDM (1983) High-dose cyclophosphamide with autologous marrow transplantation as initial treatment of small cell carcinoma of the bronchus. Cancer Chemother Pharmacol 10: 205−207

Spitzer G, Dickie KA, Litam J, Verma DS, Zander A, Lanzotti V, Valdivieso M, McCredie KB, Samuels ML (1980) High-dose combination chemotherapy with autologous bone marrow transplantation in adult solid tumours. Cancer 45: 3975−3985

Stewart P, Buckner CD, Thomas ED, Bagley C, Bensinger W, Clift RA, Appelbaum FR, Sanders J (1983) Intensive chemoradiotherapy with autologous marrow transplantation for small cell carcinoma of the lung. Cancer Treat Rep 67: 1055−1059

Trask CL. Joannides T, Harper PG, Tobias JS, Spiro SG, Geldes DM, Souhami RL (1984) Cancer (in press)

Wolff SN, Fer MF, McKay C, Hainsworth J, Hande KR, Greco FA (1982) High-dose VP16 and autologous bone marrow transplantation for advanced malignancies − a phase I study. Proc Am Assoc Cancer Res 23: C134

Is There a Role for Immunotherapy in Small Cell Bronchogenic Carcinoma?

L. Israel

Hôpital Avicenne, 125, Route de Stalingrad, 93000 Bobigny, France

Introduction

Immunodepression is seen in cancer patients and is characterized by the inability to mount inflammatory reactions, delayed hypersensitivity, lymphocyte blastogenesis in response to lectins, etc.; it may result from various mechanisms. It has been said, for example, that cachexia due to a large tumor burden might depress immune reactions, but this cannot explain the immune depression seen in early cases. On the other hand, iatrogenic or congenital immunodepression may significantly increase cancer incidence. But the most likely and most frequent mechanism is related to the adverse effect on the immune status of the host of the tumor itself. More specifically, it may be speculated that successful spontaneous tumors are precisely the ones that are able to manipulate host immunity to their own advantage.

Some attempts at nonspecific immune stimulation have been made in recent years in small cell bronchogenic carcinoma, a tumor that is known to be frequently associated with profound depression of immune response to various stimuli. The results reported will be summarized and briefly discussed here in the light of what is now known regarding the host-tumor relationship, tumor escape mechanisms, and biologic requirements for therapeutic immune manipulation in the tumor-bearing host. It will then be made clear how naive and premature have been all the attempts made in the past at "restoration" of host immunity and why so-called immunotherapy by means of nonspecific stimulating agents had to fail. New avenues for immunologic control of solid tumors will be proposed in the last part of this chapter, which apply to small cell bronchogenic carcinoma as well as to other solid tumors.

Biology of the Host-Tumor Relationship

It is becoming increasingly clear that in the vast majority of cases immune depression in the tumor-bearing host is under the biologic control of the tumor and that extremely small effects, if any, are to be expected from the systemic use of nonspecific stimulating agents such as BCG, other bacterial products, or synthetic compounds in the presence of the tumor.

"Successful" tumors, as seen by clinicians, are those that possess the ability either to invade surrounding tissues, through the action of cancer cell proteases, or to show both local invasiveness and a tendency to metastasize at distant sites. It so happens that a simple test (Israël et al. 1982) allows us to distinguish between these two categories. Tumors unable to metastasize are associated with a large systemic increase in macrophage chemotaxis, and

Recent Results in Cancer Research. Vol. 97
© Springer-Verlag Berlin · Heidelberg 1985

the ones that are able to metastasize are associated with a profound depression of macrophage chemotaxis compared with controls. Most lung cancer cases, and all small cell bronchogenic carcinomas, belong in the second category.

In order to succeed, a tumor must be able to manipulate both its micro- and its macroenvironment. The microenvironment is manipulated by excretion of a large variety of biologically active substances, including peptides that reverse chemotaxis (Fauve et al. 1974; Graham and Graham 1975), complement inhibitors (Bach-Mortensen et al. 1975), prostaglandins of the thromboxane group (Lynch and Salomon 1978), and glycoproteins that mark surface antigens (Bhavanandan and Davidson 1976; Darhovsky et al. 1980). Before it can metastasize, a tumor must also be able to manipulate its macroenvironment. This is done through a large variety of mechanisms which we, among others, have investigated in detail. One is the appearance of circulating immune complexes, formed in response to large amounts of soluble antigen shed by some tumors. Circulating immune complexes are elevated in cancer patients (Heier et al. 1977) and they are known to inhibit specific reactions against the antigen they contain. In part they can both protect the cancer cell surface from immune attack and prevent the lymphocyte and macrophage surface from being in contact with the specific antigen. In addition, we have shown that large amounts of immune complexes can inhibit PHA blastogenesis and macrophage chemotaxis in a nonspecific fashion (Samak and Israël 1982).

Another mechanism of tumor manipulation of the host immune system is connected with the large amounts of acute-phase reactant that are always found in the plasma of cancer patients (Israël and Edelstein 1978). These acute-phase reactants, especially haptoglobin, fibrinogen, orosomucoid, α_1-antitrypsin, and α_2-macroglobulin, are all antiproteases and all very rich in sialic acid. Secreted by the liver in response to tumor proteases, they bond to lymphocyte and macrophage surfaces (Israël et al. 1980b) and nonspecifically block their response to immune stimulation.

A third mechanism is represented by the excess of suppressor cells. Several studies have shown such an excess in cancer patients, and we ourselves have shown that this is so not only in the circulating blood, but also within tumors (Fujimoto et al. 1976; Samak et al. 1982). It should be remembered at this point that an effective way of eliciting suppressor cells in the experimental animal is to inject natural or artificial immune complexes.

A fourth mechanism seems to be represented by the ability of some tumors, through substances that have not yet been identified, to induce a decrease in helper T cells, a situation we have found in the draining lymph nodes of lung tumors (Samak et al. 1982). It appears, then, that tumors with metastatic potential are able to change their micro- and their macroenvironment to resist immune attack and to paralyze immune reactions of the host at various points in the immune pathway. The immune system of cancer patients is not destroyed − since, as will be shown later it always can be restored, at least for short periods, by appropriate manipulations − but it is strongly inhibited, and at the same time protected from contact with substances that might stimulate it, such as tumor-associated antigens, non-tumor antigens, and nonspecific stimulating substances. It follows that attempts at nonspecific stimulation are very likely to fail, except possibly for very short periods of time. As we will show, they have indeed failed, and other, biologically sound strategies must be found.

Various Situation for "Immunotherapy"

It should be stressed that all attempts at immunotherapy of small cell bronchogenic carcinoma have been included in a single category, namely systemic introduction of the stimulating agent in a tumor-bearing host. Though there was no other way to use nonspecific stimulating agents, it so happens that this kind of immunotherapy is unlikely to be effective. We have seen that the presence of the tumor itself induces various mechanisms of immune depression or immune inhibition of the host.

When the tumor is removed, there is very often a rise in the results of the tests used to assess the immune status. There is presumably a connection between the amount of tumor tissue and the level of immune inhibition. One can speculate that the subset of patients characterized by a metastatic potential (N+, low macrophage chemotaxis) and subjected to radical surgery might be the one in which systemic nonspecific stimulation should be tried, though with much more potent agents than have been used in the past.

Another situation is characterized by locally invasive tumors, with no metastatic potential (N−, high macrophage chemotaxis) that cannot be removed. It has been shown in several studies, both experimental and clinical, that in such cases local immunotherapy could be successful. Intratumoral injection of agents such as BCG or *C. parvum* is followed by macrophage and lymphocyte attraction, by tumor destruction, and even, as shown in experimental situations, by immunity against challenge with the same tumor.

Unfortunately small cell bronchogenic carcinoma cannot be subjected to local intratumoral immunotherapy or to surgical removal in most cases, which may explain the absence or weakness of the effects in the studies reported.

Some Results of Nonspecific Immune Stimulation in Small Cell Bronchogenic Carcinoma

Some results have indicated a positive trend and will be quoted in this section. We (Israël et al. 1977a) reported the positive effect, in a nonrandomized study, of *C. parvum* administered weekly together with chemotherapy.

At about the same time Holoye (1977) found an advantage for BCG in disseminated disease only in a study with historical controls. In a randomized study, McCracken et al. (1978) found a trend towards a better response rate ($p = 0.09$) with BCG.

Especially noteworthy is the study by Lipson et al. (1979), who randomized patients to receive thymosin or not twice weekly for 6 weeks. A significantly longer survival, without higher response rates, was found in patients given thymosin, especially in patients with higher total T cell pretreatment levels and higher pretreatment α_2HS levels.

On the other hand, however, several studies have shown no significant effect of immune stimulation in small cell bronchogenic carcinomas. MER was ineffective in the hands of Aisner and Wiernik (1980) and of Jackson et al. (1982); BCG was also shown to be ineffective by Holoye et al. (1978) and by McCracken et al. (1980), who noted an adverse effect of BCG on long-term survival.

A Conclusion on Nonspecific Immune Stimulation

If the above studies are considered, it appears that nonspecific immune stimulation is at most very marginally effective, and probably not at all, when added to chemotherapy in oat cell bronchogenic cancer. This is not unexpected in view of the mechanisms of immune escape discussed above. In the presence of a successful tumor, large amounts of circulating antiproteases coat all the immunocompetent cells and prevent them from responding to any stimulus; suppressor cells block antibody formation; and immune complexes block specific immune contact and killing. It is impossible for any nonspecific immune stimulant to even reach the cells it is supposed to stimulate. These agents may be effective against infections in immunodepressed patients, but it is impossible to see how they could be effective in such a biological situation as is established in the host by a successfully growing tumor.

Future Approaches

It follows from the above summary that any attempt at nonspecific immunostimulation − if needed at all − should only be made after manipulations directed against immune escape mechanisms. Such manipulations may involve:

a) Tumor reduction by means of chemotherapy, radiation therapy or surgery, as a means of reducing immunoinhibitory substances.
b) Removal of immune complexes by specific columns (Terman 1979).
c) Removal of immune complexes, antiproteases, and immunoregulatory peptides by repeated plasmapheresis (Israël et al. 1977b, 1980a).
d) Possibly thymectomy, as suggested by results showing both thymus revival in cancer patients and a thymus revival oriented towards suppressor cell function (Samak and Israël 1982; Israël and Samak 1983).
e) Or manipulation of thymic function by agents likely to increase T helper lymphocytes in patients who have retained some functional thymic tissue. The equipment needed for these techniques is not yet universally available, nor have many results been released; thus, these proposals are in part speculative.
f) Finally, another entirely different approach consists in the introduction of monoclonal antibody directed against possible specific antigens (Cuttita et al. 1981). Again nothing factual is yet known about this interesting possibility.

References

Aisner J, Wiernik P (1980) Chemotherapy versus chemoimmunotherapy for small cell undifferentiated carcinoma of the lung. Cancer 46: 2543−2549
Bach-Mortensen N, Osther K, Stroyer K (1975) Cl-esterase inactivators and C4 in malignant diseases. Lancet 2: 499
Bhavanandan VP, Davidson EA (1976) Characteristics of a mucintype sialoglycopeptide produced by B16 mouse melanoma cells. Biochem Biophys Res Commun 70: 139−145
Cuttita F, Rosen S, Gazdar AF (1981) Monoclonal antibodies that demonstrate specificity for several types of human lung cancer. Proc Natl Acad Sci USA 78: 4591−4595
Darhowsky D, Ziegenhagen G, Duchmann H (1980) Cl-esterase inactivation from cancer cells is C-reactive protein. Lancet 1: 149−150
Fauve RM, Hevin B, Jacob H, Gaillard JA, Jacob F (1974) Anti-inflammatory effects of murine malignant cells. Proc Natl Acad Sci USA 71: 4052
Fujimoto S, Greene MI, Sehon AH (1976) Regulation of the immune response to tumor antigens. I. Immunosuppressor cells in tumor-bearing hosts. J Immunol 116: 791−799
Israël, Samak R, Edelstein R, Bogucki D, Samak M (1980b) Immune inhibiting properties of some acute phase proteins − A possible mechanism of "immune escape" in cancer. In: Serrou B, Rosenfeld C (eds) International symposium on new trends in human immunology and cancer immunotherapy. Doin, Paris, pp 533−546
Israël L, Samak R, Edelstein R, Amouroux J, Battesti JP, de Saint-Florent G (1982) In vivo non specific macrophage chemotaxis in cancer patients and its correlation with extent of disease, regional lymph node status, and disease-free survival. Cancer Res 42: 2489−2493
Jackson DV, Paschal BR, Ferree C, Richards F, Muss HB, Cooper MR, White DR, Stuart JJ, Spurr CL, Wells B, Sartiano G, McFarland J, McCulloch J (1982) Combination chemotherapy-radiotherapy with and without the methanol-extraction residue of bacillus Calmette Guérin (MER) in

small cell carcinoma of the lung. A prospective randomized trial of the Piedmont Oncology Association. Cancer 50: 48–52

Lipson SD, Chretien PB, Makuch R, Kenady DF, Cohen MH (1979) Thymosin immunotherapy in patients with small cell carcinoma of the lung. Correlation of in vitro studies with clinical course. Cancer 43: 863–870

Lynch NR, Salomon JC (1978) Inhibition of tumor growth by aspirin and indomethacin. An examination of mechanism. Br J Cancer 38: 503

Graham RM, Graham MM (1975) Immuno-suppressive peptides in patients with cancer. N Engl J Med 292: 701

Heier HE, Carpentier N, Lange N, Lambert PH, Godal T (1977) Circulating immune complexes in patients with malignant lymphomas and solid tumors. Int J Cancer 20: 887–894

Holoye PY (1977) Chemoimmunotherapy of small cell bronchogenic carcinoma. Proc ASCO, Abstr C-45

Holoye PY, Samuels ML, Smith T, Sinkovics JG (1978) Chemoimmunotherapy of small cell bronchogenic carcinoma. Cancer 42: 34–40

Israël L, Edelstein R (1978) In vivo and in vitro studies on nonspecific blocking factors of host origin in cancer patients. Role of plasma exchange as an immunotherapeutic modality. Isr J Med Sci 14: 105–130

Israël L, Samak R (1983) Lymphocytes subpopulations in lung tumor draining lymph nodes, peripheral blood and thymic tissue at the time of resection. In: Spitzy KH, Karer K (eds) 13th International congress of chemotherapy, Vienna, September

Israël L, Depierre A, Choffel C, Milleron B, Edelstein R (1977a) Immunochemotherapy in 34 cases of oat cell carcinoma of the lung with 19 complete response. Cancer Treat Rep 61: 343–347

Israël L, Edelstein R, Mannoni P, Radot E, Greenspan EM (1977b) Plasmapheresis in patients with disseminated cancer: Clinical results and correlation with changes in serum protein. Cancer 40: 3146–3154

Israël L, Edelstein R, Samak R, Baudelot J, McDonald M, Breau JL, Mannoni P, Radot E (1980a) Clinical results of multiple plasmapheresis in patients with advanced cancer. In: Serrou B, Rosenfeld C (eds) Human cancer immunology: Immune complexes and plasma exchanges in cancer patients. Elsevier/North-Holland Biomedical Press, Amsterdam, pp 309–327

McCracken J, White J, Reed R, Livingston R, Hoogstraten B, Southwest Oncology Group (1978) Combination chemotherapy, radiotherapy, and immunotherapy for oat cell carcinoma of the lung. Proc ASCO, Abstr C-354

McCracken JD, Heilbrun L, White J, Reed R, Samson M, Saiers JH, Stephens R, Stuckey WJ, Bickers J, Livingston R (1980) Combination chemotherapy, radiotherapy, and BCG immunotherapy in extensive (metastatic) small cell carcinoma of the lung. A Southwest Oncology Group study. Cancer 46: 2335–2340

Samak R, Israël L (1982) Extraction et identification des complexes immuns circulants du sérum de cancéreux par chromatographie d'affinité suivie d'une chromatographie d'exclusion stérique à haute pression. Mise en évidence de leur effet sur la mitogenèse de lymphocytes normaux. Ann Med Interne 133: 362–366

Samak R, Israël L, de Saint-Florent G, Amouroux J, Kemeny JL, Battesti JP (1982) Caractérisation par les anticorps monoclonaux des sous populations fonctionnelles de lymphocytes présents dans les tumeurs, les ganglions régionaux, le sang périphérique et le thymus de patients porteurs de cancers pulmonaires opérables. Bull Cancer 69: 354

Terman DS (1979) Extensive necrosis of primary and metastatic canine breast adenocarcinoma (BA) after extracorporeal perfusions over *Staphylococcus aureus* columns. Clin Res 27: 392

Lung Cancer

Editor: **W. Duncan**
1984. 23 figures, 42 tables. IX, 132 pages
(Recent Results in Cancer Research, Volume 92)
ISBN 3-540-13116-7

Lung Cancer is a timely account of the clinical and social problems caused by excess cigarette smoking. Written by leading health care experts from the United Kingdom and Scandinavia, it provides essential clinical information for the recognition and management of lung cancer, while highlighting the need for allocating greater resources to counter-balance commercial promotion of cigarette consumption, particularly in the developing countries.
The book opens with a detailed description of the pathological classification of lung cancers and the techniques available for their radiological assessment. The authors then consider relapse patterns in lung cancer patients and their implications for treatment and review the most effective chemotherapeutic, radiotherapeutic and surgical methods for managing lung cancers.
In the closing chapter, attention is devoted to the future prospects for lung cancer research and treatment, emphasizing thereby the importance of international collaboration in clinical drug therapy trials and in the evaluation of new cytotoxic agents.

Early Detection and Localization of Lung Tumors in High Risk Groups

Editor: **P. R. Band**
1982. 79 figures, 66 tables. XII, 190 pages
(Recent Results in Cancer Research, Volume 82)
ISBN 3-540-11249-9

The contributions to this volume provide an authoritative review of the development, localization and detection of lung cancer in its early stages. Concentrating primarily on high risk groups such as heavy smokers and workers exposed to occupational carcinogens, they include
– a presentation of experimental aspects of tobacco smoke, radiation and asbestos carcinogenesis
– cytologic, histologic and cytochemical studies into bronchial carcinomas based on an in vivo experimental lung cancer model in dogs
– the results of new localization techniques, especially fluorescence bronchoscopy
– findings from early lung cancer detection programs in Europe, Japan and North America.
Many of the topics covered here appear for the first time in the English-language medical literature.

Springer-Verlag
Berlin
Heidelberg
New York
Tokyo

Recent Results in Cancer Research

Managing Editors:
C. Herfarth, H.-J. Senn

Springer-Verlag
Berlin
Heidelberg
New York
Tokyo

Volume 96
Adjuvant Chemotherapy of Breast Cancer
Editor: H.-J. Senn
1984. 98 figures, 91 tables. XIII, 243 pages. ISBN 3-540-13738-6

Volume 95
Spheroids in Cancer Research
Methods and Perspectives
Editors: H. Acker, J. Carlsson, R. Durand, R. M. Sutherland
1984. 83 figures, 12 tables. IX, 183 pages. ISBN 3-540-13691-6

Volume 94
Predictive Drugs Testing on Human Tumor Cells
Editors: V. Hofmann, M. E. Berens, G. Martz
1984. 87 figures, 107 tables. XII, 285 pages. ISBN 3-540-13497-2

Volume 93
Leukemia
Recent Developments in Diagnosis and Therapy
Editors: E. Thiel, S. Thierfelder
1984. 36 figures, 63 tables. IX, 305 pages. ISBN 3-540-13289-9

Volume 91
Clinical Interest of Steroid Hormone Receptors in Breast Cancer
Editors: G. Leclercq, S. Toma, R. Paridaens, J. C. Heuson
1984. 74 figures, 122 tables. XIV, 351 pages. ISBN 3-540-13042-X

Volume 90
Early Detection of Breast Cancer
Editors: S. Brünner, B. Langfeldt, P. E. Andersen
1984. 94 figures, 91 tables. XI, 214 pages. ISBN 3-540-12348-2

Volume 89
Pain in the Cancer Patient
Pathogenesis, Diagnosis and Therapy
Editors: M. Zimmermann, P. Drings, G. Wagner
1984. 67 figures, 57 tables. IX, 238 pages. ISBN 3-540-12347-4

Volume 88
Paediatric Oncology
Editor: W. Duncan
1983. 28 figures, 38 tables. X, 116 pages. ISBN 3-540-12349-0

GPSR Compliance
The European Union's (EU) General Product Safety Regulation (GPSR) is a set
of rules that requires consumer products to be safe and our obligations to
ensure this.

If you have any concerns about our products, you can contact us on

ProductSafety@springernature.com

In case Publisher is established outside the EU, the EU authorized
representative is:

Springer Nature Customer Service Center GmbH
Europaplatz 3
69115 Heidelberg, Germany